Dietary Lipids for Healthy Brain Function

Contents

Foreword

Claude Leray, a prolific and recognized scientist in lipids research, presents here his latest book in which he tackles lipids related to the neurosciences. It is a vast program! Too vast for those who are not medical doctors? No. Undoubtedly, his scientific mind, rigorous and methodical, allows him to address neuroscience holistically, covering both neurology and psychiatry. These two sides of medicine were separated a little more than 40 years ago, although so closely intertwined by the substrate on which they work the nervous system. Neuropsychiatric diseases with an aging population and the increasing stress accompanying a more urban life style are major public health concerns. The incidence of multiple sclerosis, Alzheimer's disease, Parkinson's disease, amyotrophic lateral sclerosis (ALS), and depression is constantly growing. What is the cause? Possibly aging. Also our lifestyles. Surely our environment. Neuropsychiatric medicine and research are indeed changing, and the neurosciences represent the second investment field in the development of new treatments.

A main focus of this book is to emphasize, rightly, the importance of lipids in brain function both structurally and developmentally by highlighting the current understanding and data of their key roles. It is also necessary to review the most emblematic pathologies in the neuropsychiatric field, which Claude performs with erudition and remarkable scientific caution.

You will not find here provocation, miracle recipes, or magic foods. However, you will have the most relevant data, such as the important discoveries about the role of vitamin D in the development of the nervous system, neuronal homeostasis, and neuropsychiatric diseases, all derived from fundamental and medical research. Claude stresses the importance of lipids in diseases ranging from multiple sclerosis and autism to Alzheimer's. Beyond the methodical and documented descriptions, he stresses the importance of nutrition from birth to old age to create the conditions for optimal maintenance of our neuronal assets. The key issue here is thus aging well. Clearly, lipids are not a panacea, and Claude balances the nature of his comments, supports his arguments, and justifies his conclusions.

Knowledge facilitates better insights, answers, and medical treatments; therefore, I am convinced that this work, along with its comprehensive bibliography, is important for any researcher or physician interested in the subject. It will also be a key tool for students by offering a clear overview of the topic, stimulating their curiosity and their culture, and providing an essential base to facilitate future research in the neurosciences.

Prof. William Camu
Montpellier University Hospital
Montpellier, France

Acknowledgment

I sincerely thank my friend Louis Sarliève, INSERM Research Director, for invaluable assistance in reading of the French and the English manuscripts.

I thank my wife for continuous support and invaluable involvement in the correction of the final version of the French text.

chapter one

Introduction

Research on the structure and function of the human brain started very early in the history of medical science. Greek doctors, in the sixth century BC, recognized the brain as the center of the highest human activities. At the time of Galen in the second century AD, brain anatomy was already known in detail, and the brain's importance was established as the seat of intelligence, voluntary movements, and sensations. The composition of the brain generated a significant interest as early as the time of Plato (428–348 BC), who considered this body part as equal to the bone marrow, whereas Aristotle (384–322 BC) compared the brain to a fat deposit comparable to the spermaceti found in the brain of the sperm whale.

The first observations on the fatty nature of the brain were made in the seventeenth century by the Danish physician Thomas Bartholin (1616–1680), discoverer of the lymph system and of the glands present in women that were given his name (Bartholin glands), and by the Dutch inventor of the microscope and discoverer of the spermatozoids, Antoni van Leeuwenhoek (1632–1723). But it was the early work of the French chemist Michel Eugène Chevreul (1786–1889), the founder of lipid chemistry, that led the way for other researchers in the nineteenth century to provide new knowledge about the composition of brain lipids: Nicolas Vauquelin, Jean-Pierre Couerbe, Nicolas Gobley (all in France), and Johann Thudichum (in Britain). All of these researchers described the nature of brain phospholipids, and Gobley and Thudichum also revealed the presence of few simple acids (stearic, palmitic, and oleic acids).

In the 1960s, with the advent of more efficient analytical techniques, many researchers focused on the richness of the brain and retina in docosahexaenoic acid (DHA [22:6 ω-3]), a fatty acid found in 1942 in Japanese fish (see Section 7.1). This fatty acid, called "marine fatty acid," likely appeared during the Cambrian explosion, about 600 million years ago, when its synthesis became possible because of the rising atmospheric oxygen levels above the Pasteur point responsible for aerobic life. In parallel, complex cell types also appeared that were characterized by the presence of a nucleus and several mitochondria, structures known to be common to all cellular organisms called eukaryotes.

Among the discoverers of the biochemical features characteristic of all nervous tissues, mention should also be made of John S. O'Brien (University of California–San Diego). In 1965, he was one of the first to describe accurately the fatty acid composition of several lipid fractions extracted from the white and gray substances of the human brain (O'Brien and Sampson 1965). Similar observations were made by N. C. Nielsen (Vision Research Institute, The Ohio State University) on beef retinal cells, nerve cells specialized in the perception of light (Nielsen 1979). Nielson found that the DHA content of the photoreceptor lipids was very large (36%) and higher than that measured in synaptic membranes.

Very quickly and naturally, these biochemical features led investigators to suggest that high DHA levels in brain cell membranes should correspond to a specific physiological function. One of the first assumptions was that a dietary deficiency of this fatty acid during the development of an animal or a human could hinder the formation of the myelin sheath, known to isolate nerves, thereby inducing instability in the nervous system and causing major disorders (Bernsohn and Stephanides 1967).

Subsequently and to this day, this specific affinity of the brain and retina for DHA has prompted investigations showing the plurality of functional roles for DHA in humans and animals (rat, monkey). Human clinical studies have sometimes confirmed the results found in epidemiological studies.

Despite the resistance of the nerve tissue to any change after a dietary modification, work published in 1971 first revealed that a prolonged dietary deficiency of linolenic acid (18:3 ω-3), the precursor of DHA found in plants, induced a decrease in the DHA content in rat retina (Anderson and Maude 1971). Four years later, Wheeler et al. (1975) showed that these changes were accompanied by a weakening of the electrical functioning of the retina, thereby affecting the vision of DHA-deficient animals.

Independently of this work on vision, the influence of an essential fatty acid deficiency on the general functioning of the brain was being widely explored. It seems that the subject was first described in 1966 by D. F. Caldwell in Detroit, Michigan. Caldwell and Churchill (1966) clearly demonstrated that the administration of a diet devoid of fatty acids in pregnant rats led to a serious reduction in the learning capacity of the second-generation rats. However, these investigations could not target a precise group of lipids because the food was totally delipidated.

Several researchers then showed that this lipid deficiency not only decreased the learning ability of the rats, but also induced a large drop in the DHA content in specific brain phospholipids (Lamptey and Walker 1976). Similar results were described in monkeys (Fiennes et al. 1973). In France, Bourre et al. (1989) confirmed that a diet without linolenic acid, but rich in linoleic acid, induced a significant decrease in learning ability in rats. All this research finally allowed for the determination that ω-3 fatty acids, and particularly their precursor linolenic acid, were responsible for these physiological disorders. Using a fortification of the food given to pregnant rats with a fish oil rich in DHA and eicosapentaenoic acid (EPA [20:5 ω-3]), Yonekubo et al. (1994) in Japan demonstrated an improvement of learning capacities in young rats born from these mothers, compared with animals ingesting no fish oil. In addition, this DHA and EPA intake had no effect when fed during the postpartum period. From these investigations, it can be considered that surely DHA and probably its precursor EPA are among the several components involved in the "noblest" and most vital functions of the brain.

What could be the role of these particular fatty acids of marine origin in brain function? The problem is very complex, but one of the mechanisms underlying the behavioral problems observed after a linolenic acid deficiency was proposed by Delion et al. (1994) via work done in Tours University, France. They showed that a linolenic acid deficiency was able to induce important changes in neurotransmission pathways involving dopamine- and serotonin-secreting cells in various regions of the rat brain, changes likely interfering with the animal behavior.

Work on determining the major role of ω-3 fatty acids and especially long-chain DHA in brain function, as demonstrated by Michael Crawford of the Brain Chemistry and Human Nutrition Institute in London, promoted with some success the hypothesis of their decisive intervention in the anatomical and functional development of the human brain during its development.

Indeed, from the anthropological studies we know that bipedalism, present in *Homo habilis* 2 million years ago, was contemporaneous with a significant increase in brain volume, a phenomenon likely accompanied by the adoption of a meat-rich diet. These changes were favored by the migration of this prehistoric early-human ancestor to aquatic areas rich in land animals where DHA-concentrated prey could be found. Later, about 100,000 years ago, a further increase in brain volume led to modern humans (*Homo sapiens*); this brain volume increase was contemporaneous with a new migration toward East Africa in lakeshore areas or countries close to marine environments. There, these humans found prey that supplied all of the long-chain ω-3 fatty acids (EPA, DHA) needed to build the brain. In addition, this migration was accompanied by a cultural explosion, marked by the emergence of arts, religions, and unfortunately wars. This development may ultimately be characterized more by higher brain functions than by an increased brain volume (Horrobin 1998). As emphasized by Horrobin, a prolific and popular English author, the differentiation of humans and great apes can only be based on lipid metabolism, if one considers the richness of these organic compounds in the brain and the importance of the neuronal connections.

Just as past trends have been influenced by a steady and increasing supply of DHA, it is likely that the intellectual evolution of current humans will depend on the consumption of foods rich in ω-3 fatty acids. The depletion of marine animals suitable for human consumption, as a result of intensive fishing and contamination by pesticides or heavy metals, should encourage the development of a controlled aquaculture of fish or algae producing DHA: this approach may be the only way to ensure good physical and mental health for future generations.

The role of fatty acids in the functioning of the nervous system of laboratory animals has been the subject of much research. Although a direct link has not yet been established, the effects of these fatty acids on behavior and cognitive abilities of these animals are no longer questionable. This zoopsychological approach is necessary, but the transposition of the findings from rat or even chimpanzee to the human cognitive domain remains questionable. Despite the complexity of such research, it is not surprising that neurophysiologists and psychiatrists were interested in these topics, with some of them being already investigated in animals. Much epidemiological research was recently undertaken, along with some therapeutic trials. Thus, various aspects of child development, aging, neurological disorders, or mood (or affective) disorders have been considered for therapeutic or preventive actions. Although the mechanisms involved remain poorly understood, applications of some of this research are beginning to be successfully exploited in various situations.

In addition to fatty acids, many observations have indicated that other lipids such as vitamins (A, D, and E), cholesterol, and some carotenoids could contribute to maintenance of the noble functions of the brain in aged people and also prevent serious neurological disorders such as epilepsy or multiple sclerosis.

Similarly, many mood disorders, as classified in psychopathology, seem to be under the control of these lipids. Numerous clinical studies and some experimental interventions now suggest that supplementation with some of these lipids may improve depressive and bipolar disorders, schizophrenia, autism, and attention-deficit disorders and also contribute to reduction in the intensity of aggressive or suicidal impulses.

If new results confirm the initial assumptions of the involvement of ω-3 fatty acids in brain function, and also other related compounds and vitamins belonging to the lipid group, it will be important to promote the consumption of these natural substances, the supply of which should be sufficient from a diversified diet. As suggested by D. Horrobin in 2003, the deficiencies observed in a population at risk with an unbalanced diet should be quickly filled by a supplementation with simple nutrients.

The treatment of mental disorders using a dietary approach is not yet common among the public or medical doctors. It is significant that a recent report by the Montaigne Institute in Paris evoked only the possibility of vitamin D to counteract the environmental effects on mental illness. Undoubtedly, lipid administration will gain momentum when patients realize that the current research is usually performed by government teams receiving no aid from the pharmaceutical industry. This independence may encourage a pragmatic and sympathetic consideration from the public because all of the nervous disorders described in this book could potentially be diminished without risk by a moderate dietary change or by a simple supply of an appropriate supplementation. Although this cognitive impairment approach in no way excludes modern medical therapy, patients should be aware that any alternative or at least complementary treatment already exists. This topic should be mentioned in the interview between doctor and patient, especially after taking into account the documents related to the involved problems.

These encouraging results offer the potential for a new and important treatment of many mental conditions that are currently a heavy burden on social budgets. Indeed, the World Health Organization (WHO) found that more than 450 million people suffer from behavioral or mental disorders worldwide. In the European Union, a recent analysis has shown that 27% of individuals aged 18–65 years suffered from psychiatric troubles during the past year. In France, 1 in 5 people currently suffers from a mental disorder (12 million for the whole country), compared with 1 in 10 for cancer.

The Montaigne Institute and the "FondaMental Foundation" in France estimated that in 2014 the costs associated with mental illness would reach nearly 110 billion euros per year or 5.8% of the gross domestic product. In comparison, the cost of cancer for the society was estimated at 60 billion euros and that of cardiovascular diseases at 30 billion euros. As emphasized in the Montaigne Institute report, only 2% of the budget of biomedical research is actually devoted to these problems. It is thus time to make the fight against mental diseases a public health priority. Taking into account the steady lengthening of the "total" life expectancy and the incidence of mental diseases related to old age, the main challenge of medicine in this twenty-first century is to increase the "healthy" life expectancy. Failure to achieve this ambitious goal will lead to formidable economic challenges for all nations in the management of an increasing number of frail or dependent older people.

The purpose of this book is to focus on the most important and recent work on food lipids and human health, placing the work in a historical context. Such research has provided some indisputable evidence of the beneficial effects, even for a moderate intake, of some specific lipids, sometimes absent or introduced at too low amounts in the normal diet. It is hoped that this information could be propagated widely to more easily preserve and improve brain development in young people and mental health of some adults, with only slight dietary modifications. Furthermore, these improvements involve only natural products that are much cheaper than the current traditional drugs. Despite the lack of support by pharmaceutical companies, the advances of these "nutritional treatments" as highlighted in this book may be immediately applied by healthcare personnel for the greatest benefit of patients and the global health budget.

It is regrettable that the basic rules of nutrition and dietetics are not part of the general culture, a likely consequence of the absence of education in these matters at all levels of schooling. Hopefully, the recent relationships noted between mental or neurological troubles and food lipids will spur people to take responsibility for their health status. The information in this book details how to live to old age in good health and in full autonomy, particularly by slowing the inevitable decline of the upper brain functions and by trying to avoid the development of the most disabling nervous disorders.

References

Anderson, R.E., Maude, M.B. 1971. Lipids of ocular tissues – The effects of essential fatty acid deficiency on the phospholipids of the photoreceptor membranes of rat retina. *Arch. Biochem. Biophys.* 151:270–6.

Bernsohn, J., Stephanides, L.M. 1967. Aetiology of multiple sclerosis. *Nature* 215:821–3.

Bourre, J.M., Francois, M., Youyou, A., et al. 1989. The effects of dietary alpha-linolenic acid on the composition of nerve membranes, enzymatic activity, amplitude of electrophysiological parameters, resistance to poisons and performance of learning tasks in rats. *J. Nutr.* 119:1880–92.

Caldwell, D.F., Churchill, J.A. 1966. Learning impairment in rats administered a lipid free diet during pregnancy. *Psych. Rep.* 19:99–102.

Delion, S., Chalon, S., Hérault, J., et al. 1994. Chronic dietary alpha-linolenic acid deficiency alters dopaminergic and serotoninergic neurotransmission in rats. *J. Nutr.* 124:2466–76.

Fiennes, R.N., Sinclair, A.J., Crawford, M.A. 1973. Essential fatty acid studies in primates linolenic acid requirements of capuchins. *J. Med. Primatol.* 2:155–69.

Horrobin, D.F. 1998. Schizophrenia: the illness that made us human. *Med. Hypotheses* 50:269–88.

Lamptey, M.S., Walker, B.L. 1976. A possible essential role for dietary linolenic acid in the development of the young rat. *J. Nutr.* 106:86–93.

Nielsen, N.C., Fleischer, S., McConnell, D.G. 1979. Lipid composition of bovine retinal outer segment fragments. *Biochim. Biophys. Acta* 211:10–19.

O'Brien, J.S., Sampson, E.L. 1965. Fatty acid and fatty aldehyde composition of the major brain lipids in normal human gray matter, white matter, and myelin. *J. Lipid Res.* 6:545–51.

Wheeler, T.G., Benolken, R.M., Anderson, R.E. 1975. Visual membranes: specificity of fatty acid precursors for the electrical response to illumination. *Science* 188:1312–14.

Yonekubo, A., Honda, S., Okano, M., et al. 1994. Effects of dietary fish oil during the fetal and postnatal periods on the learning ability of postnatal rats. *Biosci. Biotech. Biochem.* 58:799–801.

chapter two

Brain development

It has long been known that low-weight newborns (less than 2500 g) are more common in environments with the lowest socioeconomic status. After examining the dietary habits of mothers, it became clear that maternal nutrition plays a key role. For example, a rigorous study in East London found that mothers of such children had a dietary energy deficiency, but that deficit could be mostly attributed to lipids, a relationship implying logically a deficiency in essential fatty acids as well as lipidic vitamins such as vitamin D and E (Crawford et al. 1986). Although the hypothesis of essential fatty acid involvement was quickly confirmed by the analysis of maternal blood, the intervention of vitamins, in particular vitamin D, has still to be confirmed.

The benefits of breastfeeding in child survival were recognized long ago in that breast milk had a role in preventing the sometimes fatal effects of bacterial infections. It is now certain that it guarantees the development of intelligence, higher performance in school, and even social level in adults (Victora et al. 2015).

The influence of nutrition on brain development is in short a manifestation of neuronal plasticity as mentioned by neurologists when put at the service of rehabilitation or functional repair of brain damage. The harmonious development of the brain and cognitive performance in children as in adults is conditioned by several lipid nutrients. Unfortunately, only three of these lipids have been the subject of specific research: ω-3 fatty acids, vitamin D, and vitamin E. The effects of vitamin A are not well known because experimental research used global supplementation of various micronutrients, thereby masking the specific effects of that vitamin.

2.1 ω-3 Fatty acids

Among the essential fatty acids (Section 7.1), the links between ω-3 fatty acids and cerebral function were established early. Since the 1960s, it is known that docosahexaenoic acid (DHA [22:6 ω-3]) is the brain's major ω-3 fatty acid (about 10%–15% of total fatty acids, about 5 g in an adult brain), with the other ω-3 fatty acids amounting to less than 1% of total fatty acids. The DHA concentration varies according to the diet and the age of the subject; it is higher in the young and lower in the elderly. It is synthesized, as eicosapentaenoic acid (EPA [20:5 ω-3]), by marine phytoplankton and some animals, but not by higher plants. Notably, these two "noble" fatty acids, DHA and EPA, are absent from vegetable oils and seeds and are found at very low concentrations in fats of mammalian or poultry meat and in milk and eggs. The most generous sources are marine fish (or fish oil) and some other marine animals (molluscs, shellfish) (Section 7.2). Humans may directly absorb these fatty acids from food, and they also can synthesize them, albeit slowly, from linolenic acid (18:3 ω-3) that is present in plants and mainly in some vegetal oils (walnut, soybean, linseed). That biosynthesis, the exact efficiency of which is still controversial, is complex and involves a cascade of enzymes that sequentially elongate, desaturate, and oxidize the carbon chain (Section 7.1).

EPA and DHA biosynthesis efficiency appears to be higher in women than in men, but it has been shown that a dietary supplementation of linolenic acid in pregnant women has no effect on the DHA levels in maternal blood (de Groot et al. 2004). So, a strict vegan diet (excluding products and by-products of animal origin) could not be compatible with normal fetal development, although several observations suggest that an intake of only linolenic acid would be sufficient to maintain suitable brain DHA levels. This important issue deserves further epidemiological research. Importantly, throughout pregnancy, the placenta facilitates the transfer of DHA from the mother's body to the fetus, with its supply being ensured to the newborn through milk (breast or formula). Because DHA level in milk depends on the maternal diet, it is recommended that nursing women continue to consume foods rich in marine products. Indeed, it has been shown that even in Denmark, where fish is frequently consumed, breast milk provides only one fifth of the recommended vitamin D intake to the newborn (Streym et al. 2016).

For ethical reasons, experiments using dietary restriction of ω-3 fatty acids could be performed only in animals. Despite the usual objection for their transposition in humans, it has been established that an ω-3 fatty acid deficiency in developing animals produced a DHA depletion in the brain and that this depletion was associated with lower learning abilities. Many studies have confirmed that lower DHA levels in nerve cell membranes always induced a slowdown in neurogenesis, the formation of neuronal connections, and cell migration, with these events having negative consequences for brain growth and function.

It seems now clear that DHA deficiency in humans as in animals is critical for brain development, but details and importance of the effects are not fully understood. The impossibility to perform experiments in humans, as in rats, explains the lag time for our knowledge in this area. Moreover, in mammals the main fact that emerges is the very different timing of the brain growth spurt in relation to birth in different species (Figure 2.1). These features are at the origin of the concept of vulnerability during a so-called critical period when fast changes in function and structure occur.

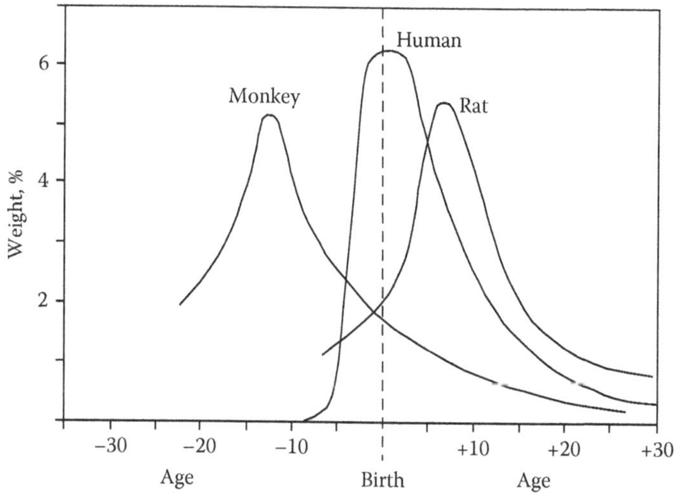

Figure 2.1 Evolution of brain growth in human, monkey, and rat. Weight increase is expressed in percentage of the brain weight in adults. The time unit has been adapted for each species: 1 day in rat, 4 days in rhesus monkey, and 1 month in human. (Modified from Dobbing, J., and Sands, J., *Early Hum. Dev.*, 311, 74–83, 1979. With permission.)

Thus, unlike in monkeys, rats are born very immature, with this feature being advantageous for the investigator. In contrast, the central nervous system of humans grows very rapidly in the perinatal period. Therefore, we must remain cautious about the conclusions drawn from experiments done in different animal species focusing on the central nervous system as well as for other physiological situations that may vary during body development.

The studies of human brain development have suggested that incorporation of DHA is essential at the end of the gestation period and immediately after birth (Clandinin et al. 1980). Indeed, it has been found that nearly 80% of the sum of embryonic arachidonic acid (20:4 ω-6) and DHA are deposited in the last 3 months of gestation. During these 3 months, the fetus has accumulated about 67 mg of DHA per day and about 75 mg during breastfeeding (Makrides and Gibson 2000). The whole amount should come from the lipid deposit of the mother, with these stocks being recovered during several months after birth. Considering these data, it is easier to appreciate the maternal needs of ω-3 and ω-6 fatty acids during pregnancy and later, especially in the case of a long-lasting breastfeeding.

Through animal studies, it can be assumed that intake of polyunsaturated fatty acids guarantees harmonious development of brain functions in the early childhood, after 2 years, and toward the final period of brain maturation.

A confirmation of this hypothesis was reported in 1990 among low-weight newborns (1000–1500 g) via work done by Dr. R. D. Uauy (Department of Pediatrics, Medical School of Dallas, Texas). Indeed, Uauy et al. (1990) demonstrated that the addition of ω-3 fatty acids from fish oil in reconstituted milk provided the same results as breast milk in relation to retinal function. The addition of linoleic acid (LA [18:2 ω-6]) added or not to linolenic acid (18:3 ω-3) provided the worst results. Subsequent research, largely carried out by Dr. D. L. O'Connor and colleagues at The Hospital for Sick Children (SickKids) in Toronto, Canada, has led to the same conclusions. In addition, there were beneficial effects on both motor development at 12 months and vocabulary comprehension at 14 months (O'Connor et al. 2001).

Other research in this area is less categorical about the favorable effects of DHA, and all neonatologists are waiting more coordinated work with greater numbers of children subjected to comparable experimental procedures and finally tested with unified tools to better assess their psychomotor development (Molloy et al. 2012). Research has identified the essential role of DHA and sometimes arachidonic acid in the anatomical and functional maturation of the brain. For example, blood DHA levels were observed to be higher in breastfed children than in those fed with a specialized newborn formula (Sanders and Naismith 1979). In rhesus monkeys (Connor et al. 1990) as well as in humans (Makrides et al. 1994), brain analyses have shown that breastfeeding ensured the highest DHA concentrations. In addition, Crawford (1993) promoted the idea that DHA was, as arachidonic acid, a vital compound in breast milk in ensuring the best development of a child's brain. Shortly afterward, a large meta-analysis consisting of 20 clinical studies conducted in six countries showed that compared with infant formulas, breastfeeding was significantly associated with better cognitive development (Anderson et al. 1999). Importantly, at that time infant formulas were not supplemented with DHA or arachidonic acid, and the ongoing changes in the production of formulas resulted from this research.

Many studies have been carried out to examine the possible effects of DHA intake on cognitive and motor development in children. In Denmark, Lauritzen et al. (2005) could not identify any effect. These negative results have been explained by the experimental use of too low amounts of DHA, an usually large consumption of fish in northern populations, or by balanced diets already rich in essential fatty acids. Moreover, these compounds may induce very early in life psychomotor effects that are unable to be detected by the current tests (Section 7.9) due to their unsuitability or low sensitivity.

The large DOMIno study, conducted in five Australian maternity hospitals, based on a supplementation of pregnant women (from the 21st week to birth) with DHA, did not detect any modification of cognitive and language development in children aged 18 months or 4 years (Makrides 2016). These negative findings seem to highlight the importance of breastfeeding because these Australian children were fed only infant formulas.

Several studies have shown that after 4 to 6 months of normal breastfeeding, a diet enriched with DHA for 1 year had a beneficial effect on the development of vision in children (Hoffman et al. 2003).

The clinical work published between 2001 and 2008 by Dr. I. B. Helland (Department of Pediatrics, University of Oslo, Norway) provides the first serious database enabling the conclusion for improved child mental development when the mother has ingested fish oil during 5 months before birth and during the first 3 months of breastfeeding. These beneficial effects are clearly observed at 4 years or age (Helland et al. 2003), and some of them still remain even after 7 years of age (Helland et al. 2008).

How were these conclusions reached? In these studies, selected mothers were given daily for defined periods 10 mL of cod liver oil (providing approximately 2.5 g of ω-3 fatty acids). Control subjects received corn oil (providing about 4.7 g of LA and 92 mg of linolenic acid). At the age of 4, the children of these mothers were submitted to intelligence testing according to the Kaufman Assessment Battery for Children. This classic multi-subtest battery (Section 7.10) aims to evaluate both intelligence and knowledge by using three scales: sequential processing, simultaneous processing, and nonverbal abilities. The study of these 4-year-old children has shown that the children from mothers ingesting fish oil (48 children) tested 4.1 intelligence quotient (IQ) points higher than those from mothers ingesting vegetable oil (36 children). The former children had also a head circumference at birth that was significantly higher than that of the latter children. At the age of 7 years, the researchers did not find any IQ difference between the two groups; however, the positioning on the scale of the sequential processing was still in favor of children born from mothers who were supplemented with fish oil.

As the authors pointed out with some humor, if new teaching methods could increase the IQ by 4 points, education authorities would immediately implement the methods. So, why are these intelligence effects in children via dietary supplementation administered in pregnant women ignored? In addition, such treatment is without side effects. If there is any doubt, why not extend these experiences to several groups of women belonging to different social positions? The interest of such results would justify the cost of the research work needed to obtain them.

Another study, published by Dr. E. E. Birch (University of Texas–Dallas) and colleagues, underscores the importance of a diet enriched with DHA (together with arachidonic acid) for only 17 weeks after birth on visual acuity and cognitive development (Birch et al. 2007). By comparing the effects of breastfeeding and infant formula supplemented or not with DHA plus arachidonic acid, Birch et al. (2007) clearly showed that the worst results on visual acuity and IQ based on verbal expression are observed among children receiving no DHA after 4 years. A large European study (United Kingdom, Belgium, Italy) has revealed that 6-year-old children who had received a diet enriched with DHA and arachidonic acid for 4 months after birth were faster at processing information compared with children who received an unsupplemented formula (Willatts et al. 2013).

A review conducted by Protzko et al. (2013) at the University of New York focused on a dozen recent studies exploring the effects of a maternal supplementation with at least 1 g of DHA or an infant supplementation with a reconstituted milk containing up to 0.5 g of DHA and EPA per 100 g of lipids. All the screened studies have shown that such a nutritional therapy clearly enabled the measurement of higher IQ several years later. After such results, Dr. Protzko did not hesitate to give as a title for his article, "How to make a young child smarter: evidence from the database of raising intelligence."

The scientific community has not entirely adopted the providential efficiency of ω-3 fatty acids in improving child neurological development. Those who are the most critical will hide behind the relatively inconclusive conclusions that arose from some meta-analyses, such as the analysis performed in Australia (Gould et al. 2013). Studies have highlighted many methodological limitations, but despite reservations, a benefit of the supplementation with ω-3 fatty acids on cognitive development in 2- to 5-year-old children has been validated. The complexity of this research is noticeable when one considers the work of Dr. C. L. Jensen and co-workers (Baylor College of Medicine, Houston, Texas), performed with 5-year-old children whose mothers had received 200 mg of DHA daily for the first 4 months of breastfeeding (Jensen et al. 2010). The results have shown that among the battery of 12 neuropsychological tests, only the test measuring the children's attention revealed a highly significant positive effect of the maternal treatment.

In another study, the offspring of an important cohort of 338 women in Mexico participating in a trial of daily 400-mg DHA supplementation during the latter half of pregnancy were assessed (Ramakrishnan et al. 2015). At 18 months of age, no overall differences in infant cognitive, motor, or behavioral development, as measured by the Bayley Scales of Infant Development (Section 7.9), were found. Nevertheless, the supplementation has attenuated the positive association between home environment and psychomotor development index observed in controls, suggesting potential benefits for children living in poor-quality home environments. The follow-up of the same children up to 5 years of age showed that DHA supplementation in the second half of pregnancy had a significant potential to improve sustained attention in preschool children (Ramakrishnan et al. 2016).

A follow-up of the large DOMInO trial is presently being carried out to explore the effects of prenatal DHA supplements on child development beyond the age of 3 years; the results are expected ca. 2018 (Gould et al. 2016).

It is clear that research of the effects of some nutrients on cognitive development is paved with many difficulties, thereby explaining the wide dispersion of results and the complexity of the comparisons between investigations that may seem, at first, similar. The greatest challenges remain the need to select subjects with well-known fatty acid status; to design well-defined control groups; and to do experiments for prolonged times, with the time needed to feed the brain with lipids not being that of the general body feeding.

The determination of the organizers of the large French EDEN survey was appreciated in conducting cooperative programs in several epidemiology laboratories of Institut National de la Santé et de la Recherche Médicale (INSERM) and in the university hospitals of Poitiers and Nancy (https://eden.vjf.inserm.fr/). That survey consists of a general study of many people focusing on what determines child psychomotor development and health before and after birth.

Among several topics, nutritional intake of ω-3 and ω-6 fatty acids was estimated in late pregnancy and was associated with the nursing times, with evaluations of language at the age of 2 years and psychomotor development at the age of 3 years made by parents and assessments made by psychologists at the age of 3 years.

Taking into account a set of characteristics in children and their families, the first results showed that the longer the breastfeeding, the better the children's cognitive performances at the ages of 2 and 3 years (Bernard et al. 2013). Among nonbreastfed children, the lower the ω-6 fatty acid/ω-3 fatty acid ratio in the maternal diet, the higher the psychomotor development scores in children at the ages of 2 and 3 years. Moreover, even if mothers have a diet rich in ω-6 fatty acids in late pregnancy, better language development was observed in children at the age of 2 years when breastfeeding was practiced longer.

Ethically acceptable ways to estimate the importance of arachidonic acid in the presence of DHA include the comparison of dietary treatments with DHA only and with DHA and arachidonic acid or the comparison of treatments with various ω-6/ω-3 fatty acid ratios. The latter option was adopted in a study of very preterm infants (<1500 g at birth) comparing two levels of arachidonic acid with constant DHA. The outcomes were that infants consuming formulas with greater arachidonic acid (twofold) had better psychomotor development at 2 years of age (Alshweki et al. 2015). Furthermore, the psychomotor development of the former group was similar to that of comparable infants who were fed exclusively with breast milk. As emphasized by the authors, these results may be explained if it is recalled that arachidonic acid is a key component of cell membranes, serving as a precursor to prostaglandin formation, and being involved in the signaling systems of the brain. Generally, the weight of existing clinical evidence favors arachidonic acid inclusion with DHA.

Therefore, these important findings suggest that the ω-6/ω-3 fatty acid ratio in the maternal diet may directly influence the development of children's brains, mainly during pregnancy, but that breastfeeding can still overcome nutritional mistakes of the mother, at least for the quality of lipids.

In the future, the monitoring of children up to the age of 5 years will enable clinicians to examine the persistence of the beneficial effects of a steady ω-3 fatty acid intake.

These early results showing all the beneficial effects in newborns of a natural diet "for their growth, development and short, medium and long term health" echo the international recommendations made by WHO in 2007 advocating an exclusive breastfeeding up to 6 months (http://whqlibdoc.who.int/publications/2008/9789241596664_eng.pdf).

It does not seem possible to doubt the long-term effects when taking into account the results obtained in Brazil from a study of about 3500 newborns (Victora et al. 2015). This study has indeed verified, on the basis of the IQ estimated between 6 months and 30 years after birth, that breastfeeding improves intellectual performance. That improvement may have a significant impact detectable 30 years later by a higher level of education and even a higher income in adulthood. Victora et al. (2015) determined that children breastfed for 1 year have an IQ that is 4 points higher than that measured in children breastfed only 1 month. These results therefore confirm the findings reported by the WHO after an analysis of 14 studies published from 1988 to 2011: "This meta-analysis suggests that breastfeeding is associated with increased performance in intelligence tests in childhood and adolescence, of 3.5 points on average" (Horta and Victora 2013). Many studies of brain imaging have confirmed these findings, stating that breast milk allows the brain to develop faster. In addition, at Brown University (Providence, Rhode Island), Deoni et al. (2013) established that breastfeeding promotes myelin development, especially in brain areas related to language, emotions, and cognitive abilities. Also, a 7-year longitudinal study in Australian preterm infants determined that a predominantly breast milk feeding in the first 28 days of life was associated with greater gray matter volume and better IQ, academic achievement, working memory, and motor function at 7 years old (Belfort et al. 2016).

It is surprising that the recommendations made in the early twenty-first century correspond to the conclusions made almost a century ago in the United States after a study of a "socially disadvantaged group" (Hoefer and Hardy 1929). In France, the National Program for Nutrition and Health has recommended breastfeeding "exclusively up to 6 months age and at least up to 4 months age for health benefits" (Hercberg et al. 2008). In addition, to the increase in the breastfeeding frequency from birth, that program recommended to increase its duration, if possible over 6 months, even during food diversification, a time when foods and beverages other than milk are introduced. Similarly, WHO recommended that breastfeeding be initiated within the first hour of birth and be exclusive for 6 months, with the introduction of complementary food after 6 months and continued breastfeeding up until 2 years or beyond.

Despite these common sense recommendations that are also based on corresponding findings in a multitude of scientific work, postnatal feeding, although improving in all countries, is still far from matching recommendations of the official texts.

In the United States, according to the Department of Health and Human Services, breastfeeding rates continue to rise. In 2011, 79% of newborn infants started to breastfeed, but breastfeeding did not continue for as long as recommended. Of infants born in 2011, only 49% were breastfeeding at 6 months and 27% at 12 months.

In Europe, WHO has estimated that only 25% of infants were exclusively breastfed for the first 6 months during a 2006–2012 study, compared with 43% in South East Asia. WHO has recommended that breastfeeding be initiated within the first hour of birth and be exclusive for 6 months, with the introduction of complementary food after 6 months and continued breastfeeding up until 2 years or beyond. In Europe, the country with the highest rate of breastfed babies is Norway: 99% of new mothers initiate breastfeeding at the hospital and 80% still do it after 6 months.

In France, the Epifane 2012–2013 study (Perinatal and Nutritional Monitoring Unit, Institute of Health Monitoring, Uspen) revealed that at birth 59% of infants are breastfed. However, 3 months later, no more than 39% were breastfed, with only 10% exclusively. After 6 months, 23% of children were still breastfed, but only 1.5% exclusively. Thus, the median value of breastfeeding is 15 weeks and that of the exclusive breastfeeding is only 24 days. As noted by the authors of that report, it seems imperative to better spread and adapt the messages on infant feeding during the first year of life.

To avoid the problems associated with analytical costs, and lengthy and cumbersome food surveys, other investigators have used an "ecological" approach, already adopted in various epidemiological studies such as those aiming at the determination of the incidence of disease in different geographical regions. This compelling approach was adopted by Dr. W. D. Lassek at the University of Pittsburgh, Pennsylvania. This work (Lassek and Gaulin 2014) has taken into account the national data published in 28 countries worldwide on the fatty acid composition of breast milk and the intellectual scores of students at the end of the compulsory education (at the age of 15 years) (Figure 2.2). These data were measured

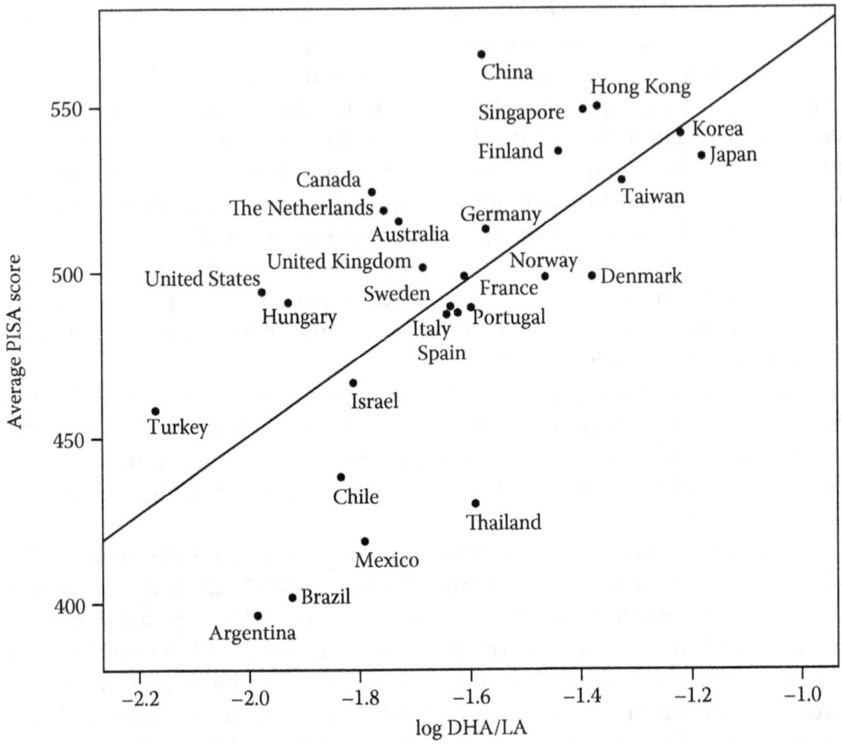

***Figure* 2.2** Relationship between the PISA test score in 15-year-old children and the docosahexaenoic/linoleic acid (DHA/LA) ratio in breast milk for 28 countries. 1: Argentina, 2: Australia, 3: Brazil, 4: Canada, 5: Chili, 6: China, 7: Denmark, 8: Finland, 9: France, 10: Germany, 11: Hong Kong, 12: Hungary, 13: Israel, 14: Italy, 15: Japan, 16: Korea, 17: Mexico, 18: The Netherlands, 19: Norway, 20: Portugal, 21: Singapore, 22: Spain, 23: Sweden, 24: Taiwan, 25: Thailand, 26: Turkey, 27: United Kingdom, 28: United States. (Modified from Lassek W.D., and Gaulin, S.J., *Prostaglandins Leukot. Essent. Fatty Acids*, 9, 195–201, 2014. With permission.)

by the Program for International Student Assessment (PISA). Notably, PISA, managed by the Organisation for Economic Co-operation and Development, aims to follow every 3 years the evolution of the performance of education systems in member countries and 60 partner countries. The results are also widely disseminated and discussed in the press, with each country trying to understand and analyze its place in the general classification. The investigations of W. D. Lassek have taken into account the assessments of the three skills measured in 2009 and 2012: reading, mathematics, and science. The authors have decided to consider the quantitative ratio of the two major milk fatty acids, DHA and LA, rather than to consider only the

DHA levels. In fact, the scientific community agrees that the physiological importance of DHA closely depends on the concentration of LA, the major component of the ω-6 fatty acid series (Section 7.1).

The main result of this large survey is that a tight correlation exists between the PISA test scores in the 28 countries and the values of the DHA/LA ratio in maternal milk. It is clear from this study that the increase in cognitive performance in school-age children may be obtained not only by increasing dietary DHA levels, but also mostly by decreasing LA levels, as LA is the major fatty acid component of our modern foods. Importantly, when the social level of the children estimated through the gross domestic product and the education spending per student are taken into account in the statistical calculation, the findings were not altered. These results confirm those made in very young children and even in animals, showing that an excess of dietary LA has a negative effect on the cognitive development. Although all work in this area did not lead to the same conclusion, none brought evidence of opposite outcomes. This should be of interest to medical professionals and especially pregnant women so that they receive recommended dietary advice to optimize the motor and cognitive development of their children. It is therefore prudent to not exceed the stated recommended daily allowance for LA, as indicated by the competent French authorities (Section 7.2). As wisely highlighted by the French National Agency on Food Safety, Environment, and Workplace Security (Agence nationale de sécurité sanitaire de l'alimentation, de l'environnement et du travail [ANSES]), these values (calculated as the percentage of the child's total energy intake) must be regarded as maximum and not optimum because they are the results of studies showing that excessive LA intake is associated with a decrease in the beneficial effects of ω-3 fatty acid (Leray 2015). The Nutrition Committee of the French Pediatrics Society endorsed these conclusions in 2014 by publishing on its website a detailed report on lipid intakes in children under 3 years of age (http://www.sfpediatrie.com).

The most efficient dose of DHA to be added to the diet of pregnant or breastfeeding women is yet to be defined, mainly in relation to their eating habits. Admittedly, the pharmaceutical industry fully understood the messages published by scientists, and it is pleased that it is possible now to get infant formulas containing up to 20 mg of DHA per 100 mL of reconstituted milk. The European Union and the US Food and Drug Administration have set up milk composition standards and have also recommended production methods preserving the added essential fatty acids. In France, ANSES has proposed new values for the recommended dietary allowances (RDAs) (Section 7.2). Regarding pregnant and nursing women, the RDA for DHA (250 mg/day) and the sum DHA plus EPA (500 mg/day) "are based on epidemiological and clinical studies that have evaluated the impact of food intake on pregnancy parameters and on visual and cognitive development in young children. They are also based on the values and the arguments for adult men or women, in terms of disease prevention." For newborns and infants, the DHA concentration must be at least 10 mg/100 mL of liquid formula. For children older than 6 months, the RDA must be between 70 and 250 mg according to age.

Advertising has naturally exploited this information; a manufacturer has even used the following assertion: "breast milk contains mostly good mono- and polyunsaturated fats essential for proper brain development and vision. A polyunsaturated fatty acid deficiency could reduce the final size of the brain by 40%!"

The epidemiological investigation of Dr. J. R. Hibbeln at the National Institutes of Health (NIH) in Bethesda, Maryland, has shown that the beneficial effects of ω-3 fatty acids on child development should only be expected for a consumption of marine products exceeding 340 g/week (Hibbeln et al. 2007). According to this specialist, the possible risk generated from the presence of potential contaminants is far outweighed by the expected health benefits. If there is too strong of a reluctance or an inability to eat that amount of fish, a supplementation with fish oil or purified preparations of ω-3 fatty acids is needed.

Gynecologists and pediatricians seem to have understood the alert messages from scientists specializing in this field because they frequently prescribe dietary supplements to pregnant or breastfeeding women that offer from 160 to 300 mg of EPA plus DHA per day (in the United States: Prenatal DHA, Opti3, Coromega omega-3, Safe catch products; in France: Femibion grossesse, Gestarelle G, Gynefam, Oligobs allaitement, Sérénité grossesse).

The influence of the ω-3 fatty acids contained in our food on brain development is the subject of numerous clinical and psychological studies and gradually becomes a significant matter of importance to public health. A concerted and large effort is needed to better understand the effects of these nutrients on the brain, mainly emphasizing the nature of the mechanisms involved. Responses have been already provided on the possible role of DHA in neurotransmission and neurogenesis and even in protection against oxidation, a well-known chemical stress that tends to kill our very active nerve cells.

Because considerable evidence has accumulated to show that genetic variation has marked effects on polyunsaturated fatty acid metabolism (Glaser et al. 2011), the study of specific gene variants should now be included in all intervention trials addressing biological effects of ω-3 fatty acids on cognitive development.

The current research interest for ω-3 fatty acid supplementation to improve offspring neurodevelopment seems obvious because 10 studies in this area were listed in 2016 on the official website of the NIH Clinical Trials (http://ClinicalTrials.gov).

Although the findings are relevant, it remains yet to inform pregnant women as soon as possible about the beneficial effects of ω-3 fatty acids, especially those of marine origin, on the brain development of their babies. Present research results suggest the future of children depends on maternal diet, so is it not worth it? It is certain that the first message to spread to expectant mothers is to follow dietary recommendations that include a regular and sufficient supply of ω-3 fatty acids of vegetable (canola, walnut oil) and animal (marine fish) origin and avoid an excess of ω-6 fatty acids from sunflower, corn, or peanut oils. Similarly, the current state of research and recommendations by the official academic associations should alert parents to keep their children until the age of 3 years on an adequate supply of DHA and linolenic acid, while minimizing the linoleic acid contribution (Section 7.2).

2.2 Vitamin D

Vitamin D is a lipid that has been the target of considerable work in relation to nervous system development. That interest seems justified because it has been shown that maternal vitamin D deficiency is linked to health problems in children, affecting, in general, fetal development, skeleton formation, and particularly the respiratory system. This deficiency is worsened by a lifestyle avoiding sunshine for fear of skin cancer or through clothing habits. Animal experiments verified that a prenatal vitamin D deficiency clearly induced interference in nerve signals between neurons and could cause impaired brain function in adults. All the physiological and biochemical data have suggested that vitamin D is an important parameter to consider in studying infant development.

As for multiple sclerosis (Section 5.2.2), epilepsy (Section 5.3.2), schizophrenia (Section 6.1.3.2), and autism (Section 6.1.5.2), cognitive development during childhood seems to be dependent on the month of birth of the subjects, giving additional evidence of a possible relationship with the synthesis capacity of vitamin D by the skin under the effect of solar radiation.

It has long been recognized that children born in winter or spring are statistically heavier and larger than those born in summer or in autumn. Academically, many studies have shown that children born in summer had statistically more learning difficulties than others, with these differences being persistent even until the age of 10 or 11 years (Martin et al. 2004). These results suggest a link between cognitive development and sunshine, and thus vitamin D. Therefore, a birth in summer implies that the first and second quarter of pregnancy elapsed in winter or early spring, poorly sunny seasons and therefore critical for vitamin D synthesis. From these observations, it is possible to deduce that a significant vitamin D supply in the third trimester of pregnancy, in summer in the Northern Hemisphere, probably has no effect on child cognitive development.

The hypothesis of the influence of vitamin D through sunshine was exploited by Dr. J. J. McGrath. McGrath et al. (2006) investigated the impact of the birth season of children of approximately 50,000 women living in 11 different US sites on various anthropometric and cognitive characteristics. The tests used were the Bayley Motor Test at the age of 8 months and the Wechsler Intelligence Test at the age of 7 years (Section 7.9). The results of that large survey have shown clearly that children born in winter and spring had higher anthropometric measurements (height, weight, head circumference) and cognitive scores than children born in summer and autumn, with these score differences being still noticeable at the age of 7 years.

After analyzing these last two works, it seems logical to conclude that vitamin D is involved for optimal physical and mental conditions in children, with this vitamin being synthesized more abundantly during sun exposure of pregnant women in summer or in early fall. Furthermore, it is possible to conclude that optimal vitamin D levels are needed during the first and second trimesters of pregnancy.

Would additional research determine more clearly the critical time? Perhaps a response could come from comparing the state of vitamin D found in the maternal blood at different periods with neurocognitive data estimated in children. Despite the importance of this subject, little research has been devoted to the relationship between vitamin D deficiency in pregnant women and the delay in child cognitive development.

One group of researchers has explored vitamin D status at the end of pregnancy, and another in early or the middle of the second quarter. In the first group, Dr. C. R. Gale (University of Southampton, UK) and collaborators could not detect any correlation between vitamin D blood levels measured in the third trimester of pregnancy and cognitive functions as well as psychological health of children estimated at the age of 9 years (Gale et al. 2008). These results allowed again for the consideration that the last trimester of pregnancy is not a critical period. These findings were confirmed in Denmark (Strom et al. 2014) and in the United States (Keim et al. 2014). In China, vitamin D determinations in umbilical cord blood in parallel with the estimation of the cognitive development of children aged 16–18 months have provided equivocal results, probably as a result of a general vitamin D deficiency diagnosed in mothers (6–24 ng/mL plasma) (Zhu et al. 2015).

In the second group, Dr. A. J. Whitehouse (Perth, Australia) and associates studied 743 mother–infant pairs; vitamin D was measured in mothers in the second quarter of pregnancy, and the children were subjected to neuropsychological testing at the ages of 5 and 10 years (Whitehouse et al. 2012). Results showed that the status of the maternal vitamin D was not related to the children's emotions or behavior but that it was statistically related to their verbal and reasoning performances (Peabody Picture Vocabulary Test, Section 7.9). Thus, a maternal vitamin D deficiency during the second trimester of pregnancy increased the risk of troubles in language development that were detectable up to 5–10 years after birth. The same year, Morales et al. (2012) in Spain showed in 1820 mother–child pairs that a high vitamin D concentration measured in the second trimester of pregnancy was significantly associated with better mental and psychomotor development, estimated with the Bayley Scale (Section 7.9). Thus, mothers with a sufficient vitamin D concentration (>30 ng/mL) had children who displayed an advantage of almost 2.5 points compared to children from vitamin D–deficient mothers (<20 ng/mL).

From these investigations, it may be concluded that an optimal vitamin D intake must be maintained in pregnant women during the first months of pregnancy, with an apparent consensus being settled around a critical window in the second quarter. It seems also important to consider the experiment conducted in India by Mithal and Kalra (2014), demonstrating that it is necessary and safe to administer to pregnant women between 1000 and 2000 international units (IU)/day of vitamin D from the second trimester of pregnancy and even 4000 IU in case of deficiency. A study in new Zealand has also shown that maternal vitamin D supplementation (100,000 IU/month) in lactating women during the first 5 months of breastfeeding improved the vitamin D status of the mother and are not only safe but also may improve the vitamin D status of the breastfeeding infant and subsequently prevent a deficiency in the first 6 months of life (Wheeler et al. 2016).

Given the immaturity of the nervous system in newborns and its slow development during the following years, it is imperative to follow prophylactic measures recommended by the official organizations for fighting against vitamin D deficiency, from birth to adolescence (Section 7.4).

All the research over the past decade suggests that a vitamin D intake consistent with the recommendations made by the competent medical authorities is a necessity in pregnant women, and then in the infant, to optimize brain development, thus ensuring optimal motor or intellectual development. Results obtained in adults in other neuropsychology areas should be undertaken in pregnant or breastfeeding women to verify the impact of vitamin D on child mental health.

2.3 Vitamin E

It is now recognized that vitamin E deficiency, more frequently accompanying malnutrition, may lead to irreversible brain damage and severe cognitive impairment (Levitsky and Strupp 1995). As shown in animals, these disorders are the result of a neuronal destruction under the effect of oxidative stress that was not counteracted by vitamin E, the natural lipid antioxidant (Section 7.5). For that reason, the requirement of vitamin E has been recommended in at-risk pregnancies, and sometimes also in women with excessive smoking, to limit the harmful effects of free radicals in the fetus (Gallo et al. 2010).

Alternatively, an excess in vitamin E may also have deleterious effects on brain development, altering neuronal plasticity in important brain areas (Salucci et al. 2014).

Early research attempting to link vitamin E to cognitive functions in young children were carried out in parallel to an exploration of the appearance of cystic fibrosis in patients aged 2–36 months (Koscik et al. 2005). By studying several subjects in the control group, the authors found that children with normal vitamin E blood concentrations (>30 ng/mL) had, after psychological examination, a score about 15 points higher than children with low vitamin E concentration (<30 ng/mL). These scores have been measured using a test of cognitive skills (TSC/2) considered as equivalent to IQ.

Similar results, but of smaller amplitude, have been obtained in China after examining 120 mother–child pairs (Chen et al. 2009). The vitamin E concentration was measured in umbilical cord blood, and cognitive development was then determined in the children at the age of 2 years. The tests were specific for the estimation of four types of operation: motor behavior, language, adaptation, and social or personal behavior.

An interesting experiment conducted in Japan with children born before term (gestation period of about 26 weeks) was reported by Kitajima et al. (2015). They showed that supplementation with α-tocopherol, the main component of vitamin E (Section 7.5), for more than 6 months may induce an increase in cognitive performances (e.g., IQ) measured at an age of 8 years. Despite the small number of individuals observed in the study (34 treated and 121 controls), these results are promising. It is hoped that further investigations including maternal treatment will quickly provide new data in this complex area of relationships between antioxidants and mental development.

These results are few, but overall they indicate that it is necessary to look for "critical periods" in the life of an individual when it is essential to provide adequate amounts of lipid molecules to influence cognitive development and behavior. It is thus likely that antioxidants must be present before or shortly after birth to prevent degenerative processes in the brain in old age, a process that can no longer be slowed. Any demonstration may be made only after the continuous study of many individuals from fetal life until several years after birth. The lack of interest from pharmaceutical companies for this goal is understandable; however, it seems desirable that the public authorities support this type of research to reduce the heavy burden of mental illness in adulthood.

References

Alshweki, A., Muñuzuri, A.P., Baña, A.M., et al. 2015. Effects of different arachidonic acid supplementation on psychomotor development in very preterm infants; a randomized controlled trial. *Nutr. J.* 14:101.

Anderson, J.W., Johnstone, B.M., Remley, D.T. 1999. Breast-feeding and cognitive development: A meta-analysis. *Am. J. Clin. Nutr.* 70:525–35.

Belfort, M.B., Anderson, P.J., Nowak, V.A., et al. 2016. Breast milk feeding, brain development, and neurocognitive outcomes: A 7-year longitudinal study in infants born at less than 30 weeks' gestation. *J. Pediatr.* 177:133.e1–9.e1. DOI: 10.1016/j.jpeds.2016.06.045.

Bernard, J.Y., De Agostini, M., Forhan, A., et al. 2013. The dietary n6:n3 fatty acid ratio during pregnancy is inversely associated with child neurodevelopment in the EDEN mother-child cohort. *J. Nutr.* 143:1481–8.

Birch, E.E., Garfield, S., Castañeda, Y., et al. 2007. Visual acuity and cognitive outcomes at 4 years of age in a double-blind, randomized trial of long chain polyunsaturated fatty acid-supplemented infant formula. *Early Hum. Dev.* 83:279–84.

Chen, K., Zhang, X., Wei, X.P., et al. 2009. Antioxidant vitamin status during pregnancy in relation to cognitive development in the first two years of life. *Early Human Dev.* 85:421–7.

Clandinin, M.T., Chappell, J.E., Leong, S., et al. 1980. Intrauterine fatty acid accretion rates in human brain: Implications for fatty acid requirements. *Early Human Dev.* 4:121–9.

Connor, W.E., Neuringer, M., Lin, D.S. 1990. Dietary effects on brain fatty acid composition: the reversibility of n-3 fatty acid deficiency and turnover of docosahexaenoic acid in the brain, erythrocytes, and plasma of rhesus monkeys. *J. Lipid Res.* 31:237–47.

Crawford, M.A. 1993. The role of essential fatty acids in neural development: Implications for perinatal nutrition. *Am. J. Clin. Nutr.* 57:703S–10S.

Crawford, M.A., Doyle, W., Craft, L., et al. 1986. A comparison of food intake during pregnancy and birthweight in high and low socioeconomic groups. *Prog. Lipid Res.* 25:240–54.

de Groot, H.C., Hornstra, G., van Houwelingen, A.C., et al. 2004. Effect of alpha-linolenic acid supplementation during pregnancy on maternal and neonatal polyunsaturated fatty acid status and pregnancy outcome. *Am. J. Clin. Nutr.* 79:251–60.

Deoni, S.C., Dean, D.C., Piryatinsky, I., et al. 2013. Breastfeeding and early white matter development: A cross-sectional study. *Neuroimage* 82:77–86.

Dobbing, J., Sands, J. 1979. Comparative aspects of the brain growth spurst. *Early Hum. Dev.* 311:74–83.

Gale, C.R., Robinson, S.M., Harvey, N.C., et al. 2008. Maternal vitamin D status during pregnancy and child outcomes. *Eur. J. Clin. Nutr.* 62:68–77.

Gallo, C., Renzi, P., Loizzo, A., et al. 2010. Potential therapeutic effects of vitamin E and C on placental oxidative stress induced by nicotine: An in vitro evidence. *Open Biochem.* 4:77–82.

Glaser, C., Lattka, E., Rzehak, P., et al. 2011. Genetic variation in polyunsaturated fatty acid metabolism and its potential relevance for human development and health. *Matern. Child Nutr.* 7:27–40.

Gould, J.F., Smithers, L.G., Makrides, M. 2013. The effect of maternal omega-3 (n-3) LCPUFA supplementation during pregnancy on early childhood cognitive and visual development: a systematic review and meta-analysis of randomized controlled trials. *Am. J. Clin. Nutr.* 97:531–44.

Gould, J.F., Treyvaud, K, Yelland, L.N., et al. 2016. Does n-3 LCPUFA supplementation during pregnancy increase the IQ of children at school age? Follow-up of a randomised controlled trial. *BMJ Open* 6:e011465.

Helland, I.B., Smith, L., Blomén, B., et al. 2008. Effect of supplementing pregnant and lactating mothers with n-3 very-long-chain fatty acids on children's IQ and body mass index at 7 years of age. *Pediatrics* 122:e472–9.

Helland, I.B., Smith, L., Saarem, K., et al. 2003. Maternal supplementation with very-long- chain n-3 fatty acids during pregnancy and lactation augments children's IQ at 4 years of age. *Pediatrics* 111:e39–e44.

Hercberg, S., Chat-Yung, S., Chaulia, M. 2008. The French National Nutrition and Health Program: 2001–2006–2010. *Int. J. Public Health* 53:68–77.

Hibbeln, J.R., Davis, J.M., Steer, C., et al. 2007. Maternal seafood consumption in pregnancy and neurodevelopmental outcomes in childhood (ALSPAC study): An observational cohort study. *Lancet* 369:578–85.

Hoefer, C., Hardy, M.C. 1929. Later development of breast fed and artificially fed infants. *JAMA* 92:615–19.

Hoffman, D.R., Birch, E.E., Castañeda, Y.S., et al. 2003. Visual function in breast-fed term infants weaned to formula with or without long-chain polyunsaturates at 4 to 6 months: A randomized clinical trial. *J. Pediatr.* 142:669–77.

Horta, B., Victora, C.G. 2013. Long-term effects of breastfeeding. A systematic review. WHO, http://apps.who.int/iris/bitstream/10665/79198/1/9789241505307_eng.pdf.

Jensen, C.L., Voigt, R.G., Llorente, A.M. 2010. Effects of early maternal docosahexaenoic acid intake on neuropsychological status and visual acuity at five years of age of breast-fed term infants. *J. Pediatr.* 157:900–5.

Keim, S.A., Bodnar, L.M., Klebanoff, M.A. 2014. Maternal and cord blood 25(OH)-vitamin D concentrations in relation to child development and behaviour. *Paediatr. Perinat. Epidemiol.* 28:434–44.

Kitajima, H., Kanazawa, T., Mori, R., et al. 2015. Long-term alpha-tocopherol supplements may improve mental development in extremely low birthweight infants. *Acta Paediatr.* 104:e82–9.

Koscik, R., Lai, H.J., Laxova, A., et al. 2005. Preventing early, prolonged vitamin E deficiency: an opportunity for better cognitive outcomes via early diagnosis through neonatal screening. *J. Pediatr.* 147:S51–6.

Lassek, W.D., Gaulin, S.J. 2014. Linoleic and docosahexaenoic acids in human milk have opposite relationships with cognitive test performance in a sample of 28 countries. *Prostaglandins Leukot. Essent. Fatty Acids* 9:195–201.

Lauritzen, L., Jørgensen, M.H., Olsen, S.F. 2005. Maternal fish oil supplementation in lactation: effect on developmental outcome in breast-fed infants. *Reprod. Nutr. Dev.* 45:535–47.

Leray, C. 2015. *Lipids, nutrition and health.* Bocca Raton, FL: CRC Press.

Levitsky, D.A., Strupp, B.J. 1995. Malnutrition and the brain: Changing concepts, changing concerns. *J. Nutr.* 125:2212S–20S.

Makrides, M. 2016. Understanding the effects of docosahexaenoic acid (DHA) supplementation during pregnancy on multiple outcomes from the DOMInO trial. *OCL* 23:D105.

Makrides, M., Gibson, R. 2000. Long-chain polyunsaturated fatty acid requirements during pregnancy and lactation. *Am. J. Clin. Nutr.* 7:307S–11S.

Makrides, M., Neumann, M.A., Byard, R.W., et al. 1994. Fatty acid composition of brain, retina, and erythrocytes in breast-and formula-fed infants. *Am. J. Clin. Nutr.* 60:189–94.

Martin, R.P., Foels, P., Clanton, G., et al. 2004. Season of birth is related to child retention rates, achievement, and rate of diagnosis of specific LD. *J. Learn. Disabil.* 37:307–17.

McGrath, J.J., Saha, S., Lieberman, D.E., et al. 2006. Season of birth is associated with anthropometric and neurocognitive outcomes during infancy and childhood in a general population birth cohort. *Schizophr. Res.* 81:91–100.

Mithal, A., Kalra, S. 2014. Vitamin D supplementation in pregnancy. *Indian. J. Endocrinol. Metab.* 18:593–6.

Molloy, C., Doyle, L.W., Makrides, M., et al. 2012. Docosahexaenoic acid and visual functioning in preterm infants: A review. *Neuropsychol. Rev.* 22: 425–37.

Morales, E., Guxens, M., Llop, S., et al. 2012. Circulating 25-hydroxyvitamin D3 in pregnancy and infant neuropsychological development. *Pediatrics* 130:e913–20.

O'Connor, D.L., Hall, R., Adamkin, D., et al. 2001. Growth and development in preterm infants fed long-chain polyunsaturated fatty acids: A prospective, randomized controlled trial. *Pediatrics* 108:359–71.

Protzko, J., Aronson, J., Blair, C. 2013. How to make a young child smarter: Evidence from the database of raising intelligence. *Perspect. Psychol. Sci.* 8: 25–40.

Ramakrishnan, U., Gonzalez-Casanova, I., Schnaas, L., et al. 2016. Prenatal supplementation with DHA improves attention at 5 y of age: A randomized controlled trial. *Am. J. Clin. Nutr.* 104:1075–82. DOI: 10.3945/ajcn.114.101071.

Ramakrishnan, U., Stinger, A., DiGirolamo, A.M., et al. 2015. Prenatal docosahexaenoic acid supplementation and offspring development at 18 months: Randomized controlled trial. *PLoS One* 10:e0120065.

Salucci, S., Ambrogini, P., Lattanzi, D., et al. 2014. Maternal dietary loads of alpha-tocopherol increase synapse density and glial synaptic coverage in the hippocampus of adult offspring. *Eur. J. Histochem.* 58:2355.

Sanders, T.A., Naismith, D.J. 1979. A comparison of the influence of breast-feeding and bottle-feeding on the fatty acid composition of the erythrocytes. *Br. J. Nutr.* 41:619–23.

Streym, S.V., Højskov, C.S., Møller, U.K., et al. 2016. Vitamin D content in human breast milk: A 9-mo follow-up study. *Am. J. Clin. Nutr.* 103:107–14.

Strom, M., Halldorsson, T.I., Hansen, S., et al. 2014. Vitamin D measured in maternal serum and offspring neurodevelopmental outcomes: A prospective study with long-term follow-up. *Ann. Nutr. Metab.* 64:254–61.

Uauy, R.D., Birch, D.G., Birch, E.E., et al. 1990. Effect of dietary omega-3 fatty acids on retinal function of very-low-birth-weight neonates. *Pediatr. Res.* 28: 485–92.

Victora, C.G., Horta, B.L., Loret de Molaet, C., et al. 2015. Association between breastfeeding and intelligence, educational attainment, and income at 30 years of age a prospective birth cohort study from Brazil. *Lancet Glob. Health* 3:e199–205.

Wheeler, B.J., Taylor, B.J., Herbison, P., et al. 2016. High-dose monthly maternal cholecalciferol supplementation during breastfeeding affects maternal and infant vitamin D Status at 5 months postpartum: A randomized controlled trial. *J. Nutr.* 146:1999–2006. DOI: 10.3945/jn.116.236679.

Whitehouse, A.J., Holt, B.J., Serralha, M., et al. 2012. Maternal serum vitamin D levels during pregnancy and offspring neurocognitive development. *Pediatrics* 129:485–93.

Willatts, P., Forsyth, S., Agostoni, C., et al. 2013. Effects of long-chain PUFA supplementation in infant formula on cognitive function in later childhood. *Am. J. Clin. Nutr.* 98:536–42S.

Zhu, P., Tong, S.L., Hao, J.H., et al. 2015. Cord blood vitamin D and neurocognitive development are nonlinearly related in toddlers. *J. Nutr.* 145:1232–8.

chapter three

Cognitive development

Neurobiologists have long established that the interactions necessary to develop both the organization and the function of the brain are present in a complex environment, mainly organic but also "sociocognitive." During childhood and adolescence, the brain continues its organization and neurobiological adaptation to the surrounding world under the control of endogenous and exogenous influences essential for regular development. The question once asked whether cognitive functions are genetically determined is no longer relevant; now, there must be added the questions of when and how are they influenced by the environment and more precisely by nutrients.

For humans, the duration of brain development is particularly long. Indeed, brain size increases nearly five times from birth to adulthood, and if the formation of the nervous system is fast at the beginning, the organization of synaptic connections and myelin sheaths is not complete until about 15–20 years. Similarly, some areas such as the prefrontal cortex and the basal ganglia reach maturity at beyond 15 years and can even continue to change throughout life.

Therefore, this slow brain maturation, mainly based on the addition and remodeling of anatomical structures, may be modulated by lipid nutrients, essential compounds for building cell membranes and synthesis of several neurotransmitters. Thus, even in adults, the brain gifted with plasticity and depending on food and environment will evolve toward a progressive development of its cognitive abilities and will decline more or less rapidly in disclosing various disorders during aging.

Among the lipid components of the human diet that may influence cognitive development, the most advanced research has targeted ω-3 fatty acids (Section 3.1); phospholipids (Section 3.2), in particular phosphatidylcholine and phosphatidylserine; and vitamin D (Section 3.3).

3.1 ω-3 Fatty acids

3.1.1 Epidemiological investigations

The results presented in Chapter 2 related to the effects of ω-3 fatty acids carried by blood to the fetus or by milk during breastfeeding on brain development have naturally led clinicians to undertake similar investigations in young children, adults, and the elderly.

The first indications of a possible link between these essential fatty acids and brain function were reported in 1982 when Dr. R. Holman described for the first time the effects of linolenic acid (18:3 ω-3) deficiency in a young girl of 6 years old (Holman et al. 1982). This famous American biochemist, and originator of the expression omega-3 or ω-3, noticed neurological symptoms and psychological disturbances appearing in the child after a 5-month ω-3 fatty acid deficiency. Subsequently, many studies have been done, but they could not lead to a clear understanding of the influence of these fatty acids on the cognitive capacity of the human brain, whether in young children or in adults. In 1994, at the beginning of this research, and during roughly the next 10 years, all of the meta-analyses have avoided the conclusion of any beneficial effect of these fatty acids on the improvement of cognitive performances in children older than 2 years (Eilander et al. 2007).

Most disagreements between the published reports are certainly attributable to many experimental inaccuracies and to the use of different neuropsychological tests. Thus, clinicians use comprehensive tests exploring various cognitive functions, such as the battery of tests according to Kaufman for children 3–18 years old, to Wechsler for 6–16 years old, or to Bayley from a few months to 3 years old (Sections 7.9–7.10). Other tests may be used to target more specific functions such as language, communication, recognition, and problem-solving. The number of tests currently available is very important and increases every year; 18 tests using computers were already listed by 2008 (Wild et al. 2008).

Moreover, in these experiments the participants consumed for different times various amounts of foods with qualitative properties validated by biochemical analyses. These cross-sectional studies are the most commonly used in an overall population (or a representative sample or cohort) and at a given time. They thus provide a snapshot image of the population studied. The results obtained by that technical approach are sometimes considered approximate or inconsistent, even when positive trends are detected. To improve the accuracy of the results, researchers have adopted more specific procedures using numerous and homogeneous groups and analytical methods less prone to criticism.

Investigators tackling these difficult psychophysiological problems frequently try to appreciate the main cognitive functions in children and adults by recognized methods. Thereafter, they compare these quantitative results to the amounts of fatty acids ingested one or more days before having a neuropsychological test(s). Because hundreds of data must be acquired, these explorations are long and expensive, but they are essential for the precise knowledge of the relationships between nutrients and brain function.

Determining cognitive performances in selected individuals is usually directed at various aspects such as perception, language, memory, reasoning, and movement, all clarifying the operation of different types of memory that are necessary for a broad knowledge. The determination of amounts of ingested fatty acids is carried out using a questionnaire on the type and weight of the foods consumed some time before. The estimated amounts of ingested fatty acids are then calculated from these weights with the aid of official food composition tables.

Among the most compelling recent explorations is that of Dr. W. D. Lassek (University of Pittsburgh, Pennsylvania), who involved more than 4000 children (equal number for each sex) aged 6 to 16 years and diverse ethnic groups (Lassek and Gaullin 2011). In all selected subjects, four cognitive tests were used: mathematics, reading, and estimation of spatial perception and numbers memory (Wechsler Test). The average scores for these four tests were taken as a measure of the cognitive performance of the children studied.

After eliminating statistically all possible bias (social and biological factors, or otherwise), this work led to the following results:

- The scores of cognitive tests were positively related to the amount of ω-3 fatty acids ingested by the subjects, boys or girls. In girls, the relationship had a magnitude about two times greater than in boys, the latter seeming less sensitive to the dietary ω-3 fatty acids. According to the authors, this sex-related difference could reflect a greater need of ω-3 fatty acids in girls for a future pregnancy and breastfeeding. This aspect has only rarely been taken into account in similar studies.
- The ω-6-to-ω-3 fatty acid ratio in the foods was inversely related to the scores obtained for cognitive measures, but the correlation was significant only in girls. The authors suggested that the high ω-6-to-ω-3 ratio in the American diet might contribute to the relatively low ranking of American children in international testing compared to children living in countries such as Japan with lower ratios. As emphasized by Dr. Lassek, these conclusions could lead to findings of considerable potential public health significance.

To understand this research, remember that the groups of subjects are heterogeneous and that the assessment of the average food intake from each subject gives an approximate value. Is the knowledge of the meals the day before a correct estimate of the nutritional status in the long term? Many clinicians doubt it. The cost of several food surveys for years would obviously be prohibitive.

Moreover, it is likely that the high significance of the ω-3 fatty acid effects reported in the previous study is related to the low consumption of these fatty acids by American children and to their excessive ω-6 fatty acid intake interfering with the ω-3 fatty acid effects. These characteristics could explain the unconvincing diversity of the results reported in several studies carried out in various regions and with subjects selected in different sociological categories.

It can also be speculated that improving the ω-3 fatty acid status could be more effective on cognitive performances in studying very young children during the period of the highest brain growth.

The great difficulty of such investigations is now clearly emerging. As highlighted by Dr. Lassek, the dietary situation of the subjects enrolled in that major study is probably related to the low score level obtained by the American children in international rankings (National Center for Education Statistics, Washington, DC). If that relationship is confirmed by similar studies in different countries and social groups, it would be possible to better appreciate the potential capacity of governments to improve the intellectual status of children going to school.

Interestingly, a further study by Dr. K. W. Sheppard (University of Carolina North, Chapel Hill) has confirmed these results in children aged 7–9 years after recording the meal composition three times per week before psychometric testing (Sheppard and Cheatham 2013). In that experiment, the most significant results were obtained with a spatial ability test, a subset of the Cambridge Neuropsychological Test Assessment Battery providing information on the contribution of the brain frontal lobes. In 2012, Prof. R. de Groot (Open Universiteit, Heerlen, the Netherlands) had already shown that among students aged from 12 to 18 years, fish consumption was positively associated with knowledge acquisition in three key disciplines. Notably, some adverse effects were detected for high fish intake and were probably related to the presence of toxic substances in that kind of food (de Groot et al. 2012).

A study by Dr. C. L. Baym (University of Illinois, Urbana) has offered a new perspective on the relationships between ω-3 fatty acids and a specific type of memory, relational memory, responsible for linking facts and events (Baym et al. 2014). This ability is known to be generated in a specific area of the brain, the hippocampus, a region belonging to the limbic system and located in the medial temporal lobe. The study demonstrated that the performances of relational memory in children 7–9 years old were proportional to the intake of ω-3 fatty acids. Conversely, the amount of ingested saturated fatty acids seemed to be associated not only with reduced performances of the relational memory but also of the visual memory (i.e., memory independent of the hippocampus, but located in the right prefrontal cortex).

To overcome the inaccuracy of food surveys, some studies have applied fatty acid analysis in blood plasma or in erythrocytes (ideal because they are long-lived cells) to estimate more accurately, but indirectly, the dietary ω-3 fatty acid intake.

Among these studies, attention must be given to the research by Dr. P. Montgomery (Oxford University, UK) who enrolled 493 children aged 7–9 years with a reading level lower than the national average (Montgomery et al. 2013). All children were analyzed for their blood ω-3 fatty acid content and underwent psychometric testing to estimate their reading ability and the level of their working memory. Montgomery et al. (2013) observed that the lowest erythrocyte concentrations of docosahexaenoic acid (DHA) were closely associated to the weakest reading ability and especially to the lowest efficiency of their working memory. They concluded that the eicosapentaenoic acid (EPA) plus DHA amount in erythrocytes, already considered a reliable health marker for the cardiovascular system, may now be used for the prediction of behavior and mental health in young children.

In addition, the research team of Prof. R. de Groot has confirmed these results, indicating that in subjects aged 13–15 years, the highest values of the "ω-3 Index" (Section 7.1) measured in the blood were correlated to better scores estimated with several psychometric tests. The most significant outcome was the observation of a positive relationship with the speed of information processing and a negative relationship with the impulsivity of the subject (van der Wurff et al. 2016a). To supplement these investigations, it is important to take into account research by the same team focusing on school performance instead of general cognition. Thus, it seems noteworthy that at age of 7, in 150 children, significant associations between DHA level measured in plasma phospholipids and both reading and spelling were found (van der Wurff et al. 2016b). Unexpectedly, it was also shown that maternal plasma DHA levels were negatively associated with arithmetic scores of children at age 7 years. No biochemical explanation for these outcomes could be evoked, but as emphasized by authors, one could speculate that this is because these different skills are located in different brain regions. Consequently, this observational study requires prudence when considering DHA supplementation during pregnancy.

Studies based on adult subjects aged between 20 and 60 years are scarce in comparison with studies with older subjects suffering from cognitive decline. Most observations using dietary surveys have led to the conclusion that there is an absence or a weak correlation between cognitive performances and DHA intake (e.g., Joffre et al. 2014). A study using blood tests has shown no effect or even a negative relationship between DHA erythrocyte concentration and learning speed (de Groot et al. 2007). Particularly, these authors selected young women about 30 years old who did not consume more than one fish serving per week.

Similarly, an original work (Johnston et al. 2013) was done by the medical services of the US Army after enrolling soldiers aged 20–54 years that were deployed during Iraq operations. The authors observed that the levels of total blood ω-3 fatty acids, estimated by the ω-3 Index (Section 7.1), were directly associated with better cognitive performances (flexibility, executive functions). The results have been even more convincing in subjects suffering sleep problems. It is easy to understand the value of such studies as part of a program of improving cognitive performances in a theater operation. Moreover, the US army is undertaking investigations to check whether a regular supplementation with ω-3 fatty acids may be beneficial for that category of personnel.

Regarding children, it seems evident that the prevention of school failures could include, without excessive costs, a dietary analysis leading when necessary to an increase of their daily intake of essential fatty acids including ω-3 fatty acids, while ensuring a control of their ω-6 fatty acid intake. These contributions could be exercised, for example, through the menu planning in school cafeterias, without neglecting simple advice to families for dining at home. The financial investment in this area would certainly be very profitable for the future health of schoolchildren and students generations. It is likely that the results would be most effective if these measures would be taken earlier in school years.

All these observations enlarge the set of data strongly suggesting that the dietary intake of ω-3 fatty acids in children consistent with official recommendations is beneficial to the acquisition and preservation of a "normal memory," even though all aspects have not yet been fully explored. Although the intake of these fatty acids has been proven effective in pregnant or nursing women or in neonates (Section 2.1), what happens in young children, especially if their nutritional status is unfavorable? Research involving nutritional interventions is thus needed to elucidate all possible effects of a change in the fatty acid status on cognitive performance.

3.1.2 *Intervention studies*

Supplementation experiments are obviously more difficult to organize than the epidemiological surveys described above; however, several teams have examined the possible beneficial effects of an ω-3 fatty acid supplementation in children, mainly if they displayed a current nutritional deficiency.

But how can we correct mental deficiencies or improve a mental state even considered "normal" with an ω-3 fatty acid fortification? Many studies have tried to answer these questions, so it seems necessary to analyze the most important, although few having ended in failure.

Below are some of the more recent experiments that have led to equivocal conclusions:

- In 2009 in Bangalore, India, 600 children aged 6–10 years were supplemented daily with or not supplemented with 100 mg of DHA and 930 mg of linolenic acid (Muthayya et al. 2009). No effect on cognitive performance could be detected after a 9-month supplementation.

- In 2010 in Newport, UK, 500 children aged 8–10 years were supplemented daily or not supplemented for 4 months with 200 mg of DHA and 28 mg of EPA (Kirby et al. 2010). No convincing neuropsychological outcomes could be detected.
- In 2012 in Oregon, Dr. J. E. Karr administered fish oil daily (480 mg DHA and 720 mg EPA) for 4 weeks to students about 20 years old (Karr et al. 2012). No cognitive benefits could be noted.

Fortunately, for the future of treatments with these compounds, now called "nutraceuticals," several studies have reached less pessimistic conclusions likely to be considered by the general population. It may be assumed that these results were obtained by better equipped investigators who can appreciate subtle changes in cognitive functions.

Below are some experiments that led to definite conclusions, otherwise indisputable, on various physiological and neuropsychological aspects:

- In 2010, one of the most comprehensive studies was conducted by a specialist of these questions, Dr. R. K. McNamara of the University of Cincinnati, Ohio (McNamara et al. 2010). Boys between 8 and 10 years old took 400 or 1200 mg of DHA/day or a placebo. After an 8-week treatment, the erythrocyte DHA was analyzed and hemodynamic changes related to brain activity were monitored by functional magnetic resonance imaging (fMRI). For the first time, it could be shown that a DHA supplementation increased the activity of a specific brain area (dorsolateral prefrontal cortex) during vigilance and attention exercises, with that change being positively correlated with the erythrocyte DHA content. This study confirmed and completed a former study done in 2005 by Dr. G. Fontani (University of Siena, Italy), showing that a fish oil supplementation (4 g/day for 35 days) was associated with an improvement in attentional and physiological functions, particularly those involving complex cortical processing (Fontani et al. 2005).
- In 2012 at the University of Oxford, UK (DOLAB study), 360 children aged from 7 to 9 years with difficulties in reading were supplemented daily with 600 mg of DHA from algal origin (Richardson et al. 2012). After a 16-week treatment, the authors found that compared with control subjects, the supplementation improved reading significantly (an advance of almost 1 month), as well as the students' behavior toward teachers. At the same time, an improvement of the working memory was detected by a team at the University of Pittsburgh, after subjects aged 18 and 25 years had taken 750 mg of DHA and 930 mg of EPA daily (Narendran et al. 2012).

The fMRI technique combined with neuropsychological tests has allowed Dr. I. Bauer (Swinburne University of Technology, Hawthorn, Australia) to conclude that in adults aged 24 years, an EPA-rich supplementation was more efficient that DHA for the participants' brains as they worked "less hard" and achieved a better cognitive performance compared to before supplementation (Bauer et al. 2014). The effect of a fish oil supplementation (1 g/day for 12 weeks) was also explored in connection with the estimation of the intensity of cerebral hemodynamics by using a noninvasive technique based on the penetration of infrared radiations through the skull (Jackson et al. 2012). Despite the lack of effect on cognitive performances in students (aged from 18 to 29 years), fish oil intake has greatly increased the amount of oxygenated blood in the prefrontal cortex while performing various cognitive tasks. These results confirm that dietary DHA probably influences brain function by modulating the bloodstream, thus neural tissue oxygenation. It is likely that the high intellectual level of the subjects and the short duration of that experiment were responsible for the lack of direct effect of the ω-3 fatty acids on cognitive functions. Curiously, the same laboratory did not detect any hemodynamic effects in the elderly (Jackson et al. 2016).

A study conducted among young aboriginal Australians aged 7 to 12 years that consumed very little fish rich in DHA and EPA also revealed that a moderate daily intake of fish oil may improve, in less than 5 months, maturity and intellectual capacity through nonverbal cognitive development (Draw-A-Person Test) (Parletta et al. 2013). A similar but more comprehensive study was undertaken in 2014 at Universidad Autónoma de Ciudad Juárez, Mexico, by Dr. V. Portillo-Reyes (Portillo-Reyes et al. 2014). Children selected in a population suffering from malnutrition were treated daily for 3 months with 180 mg of DHA and 270 mg of EPA. The use of 15 psychometric tests highlighted a significant improvement in several parameters (speed processing, visual-perceptual coordination, attention and executive functions) in more than 70% of treated children. In contrast, no memory type was improved, suggesting that it is imperative to extend in future research the range of tests and duration of treatments to better detect some specific effects of ω-3 fatty acids.

In adults ingesting low ω-3 fatty acid amounts, Dr. W. Stonehouse (University of Auckland, New Zealand) showed that a dietary supplementation with about 1 g of DHA daily for 6 months improved markedly episodic memory performances (a form of declarative memory) in women and working memory in men (Section 7.9) (Stonehouse et al. 2013). This experiment demonstrated that the results are modulated by the subject's sex (male or female); therefore, it seems necessary to better adapt the psychophysiological tests. Thus, future research could perhaps adapt the supplementation according to the subject's sex in addition to the nutritional status to improve especially one specific memory type.

Although clinical data are still limited, it seems that DHA-based treatments are more promising than those involving EPA or other ω-3 fatty acids for improving memory and learning in individuals suffering only from mild cognitive impairment (Dyall 2015). Notably, the results are, in general, more significant when subjects are not carriers of the apolipoprotein E (ApoE4) isoform, a major risk factor for several nerve diseases (Section 7.8). It is clear that in the early years of childhood there is an important maturation of the nervous system detectable at the level of neurons and their increasing number of synaptic contacts. All of these changes are accompanied by a characteristic DHA enrichment, due partially to the synthesis from precursors, such as linolenic acid, but mostly directly from the diet. The low DHA concentration in common foods consumed by children and adolescents in developed or developing countries certainly explains the effectiveness of DHA supplementation in subjects aged less than 10 years. For older subjects, several experiments have shown little effect except some changes in cerebral blood flow that are, however, important for enhancing brain metabolism. Despite the limited number of studies and the diversity of protocols, the investigators generally concluded that an EPA deficiency, DHA deficiency, or both is harmful for learning in young children.

The accuracy of the results and their significance could certainly be improved after investigations of the optimal doses of ω-3 fatty acids, the duration of clinical trials, the nutritional status of participants, or the specificity of the various psychometric tests. One of the important points to take into account seems to be a possible iron deficiency, very common in developing countries. Indeed, in children suffering from iron deficiency anemia, the administration of both iron and ω-3 fatty acids seemed to be necessary for the restoration of optimal cognitive functions (Baumgartner et al. 2015).

Since 2005, investigators have known that people are not genetically equal with regard to dietary treatments. Indeed, as noted above, the presence of different alleles of the gene encoding the synthesis of ApoE must be added as a determining factor (Section 7.8). The presence of the ApoE4 isoform has been associated with the onset, progression, and severity of many diseases, but above all, dementia or Alzheimer's disease (Huang et al. 2005). To clarify the importance of that genetic equipment, it may be simply considered that in the elderly (>64 years), healthy and without memory complaint, the amount of ω-3 fatty acids in erythrocytes was correlated with cognitive performance, but only in people lacking the ApoE4 gene (Whalley et al. 2008).

In summary, the current state of research on the relationships between dietary ω-3 fatty acids and cognitive performances in children should suggest that parents ensure a supply of these fatty acids close to the recommendations made by the medical agencies, i.e., 250 mg of EPA and DHA for children 3–9 years old and 500 mg for young people 10–18 years old (Section 7.2). Unless the child consumes one or two fatty fish servings a week, the current intake will remain low and then after medical advice the child must be regularly supplemented with ω-3 fatty acids commonly sold in capsules. The emphasis should be on the requirement to supplement children as soon as possible, and the short-term benefits will be even more evident. Parents should acknowledge rapidly the effects on educational outcomes concerning memory and attention, such as computation and reading.

3.2 Fatty acids

3.2.1 Phosphatidylcholine

The lipid phosphatidylcholine (Section 7.7.1) is ingested by humans from either meat or plants, or in the form of commercial products known as lecithins. Besides fatty acids, it represents a good source of choline, an essential substance involved in the building of cell membranes and also in many methylation reactions for the synthesis of acetylcholine, a very important neurotransmitter. The richest foods in choline are calf liver (about 400 mg/100 g) and eggs (about 250 mg/100 g). Apart from choline, phosphatidylcholine may be also a carrier of long-chain ω-3 fatty acids, as in krill oil. In that case, it will be difficult to attribute any effect to the presence of choline or to the fatty acid moieties. Despite that reservation, Japanese authors have recently awarded to ω-3 fatty acids the effects of a modest intake of krill oil (0.28 g/day of EPA + DHA) on brain function (Konagai et al. 2013). Furthermore, considering equal amounts of ω-3 fatty acids, krill oil, rich in phosphatidylcholine, proved to be more effective than a triglyceride-rich sardine oil on cognitive functions in healthy elderly.

In the 1980s, phosphatidylcholine was considered as a choline potential donor enabling the improvement of cholinergic systems that were weakened in patients suffering from dementia. It was early reported that the administration of phosphatidylcholine could improve the memory of patients suffering from Alzheimer's disease. Thereafter, many investigators explored the possible effects of the administration of this phospholipid on other cognitive functions.

Considerable research was carried out in rats with some success, either on brain development or behavior or memory. Unfortunately, little work has been done in humans, and mainly in elderly patients or patients with memory loss or dementia. Curiously, a positive result on explicit memory (or declarative memory, Section 7.9) was reported in students only 90 minutes after ingesting 25 g of phosphatidylcholine, providing 3.75 g of choline (Ladd et al. 1993). Is there a real effect or a placebo effect? It is too early to choose the explanation.

In 2003, a review by M. A. McDaniel in the United States reported some positive results, but the analysis remains questionable given the age of the patients, their nutritional history, the diversity of the selected populations, or their brain damage. A meta-analysis conducted in 2003 by the Cochrane Collaboration (UK) with 12 clinical trials involving patients with various memory impairments, but also with Alzheimer's or Parkinson's diseases, failed to demonstrate any benefit from phosphatidylcholine administration.

Some work in Japan by T. Nagata may initiate a new approach (Nagata et al. 2011). Following basic research on acetylcholine receptors in rat brain hippocampus, they obtained in humans surprisingly positive results on moderate memory loss and learning. The oral administration consisted of two types of phosphatidylcholine: one type containing palmitic and oleic acids and the other type containing only linoleic acid. The study highlighted the difficulty to draw meaningful conclusions after using complex lipids with an origin and composition poorly known, even if they are extracted from plants.

Despite uncertainty about the validity of many results, food supplements containing large amounts of plant phosphatidylcholine (soy lecithin) are available on the market with the claims of beneficial effects for memory, concentration, or sleep. These allegations concerning neurological disorders are questionable and are often based on confusion between phosphatidylcholine and phosphatidylserine, both phospholipids being present together in all natural raw extracts, but in varying proportions.

3.2.2 Phosphatidylserine

Phosphatidylserine (Section 7.7.2) is found in all plant and animal cell membranes. Its characteristic is to be located in the inner leaflet of the cell membranes where it exerts multiple functions, such as the regulation of receptors, enzymes, and ion channels. In animals, this phospholipid is mainly concentrated in the brain (about 15% of total phospholipids in humans). Many studies have shown that it plays a major role in the communication between neurons, allowing assignment to that lipid the possibility to modulate cognitive functions.

So, considering numerous encouraging findings in animals, several trials to improve memory by supplementation with phosphatidylserine have been undertaken in young children or in the elderly. The first tests were carried out in 1990 by Maggioni and co-workers in Italy. They experimented with an oral treatment with phosphatidylserine isolated from beef brain that seemed successful in improving depressive symptoms in older people as well as some cognitive parameters (Maggioni et al. 1990). Later, other researchers showed that the administration of phosphatidylserine was effective in improving memory and learning in seniors suffering from declining memory.

Early research in this area used a product extracted from beef brain, but a possible contamination by prions causing spongiform encephalopathy ("mad cow disease") has led governments to specifically prohibit the consumption and use of bovine nervous tissues. These prohibitions were initiated in 1986 in the United Kingdom and adopted in 1989 in France and in 1996 in the United States and in many other countries. Yet, it has been admitted that these animal fats should be treated by ultrafiltration and sterilization at 133°C for 20 min before their use in food products. This major crisis has led researchers to use an alternative source, if possible from plant origin, such as soy lecithin. It was therefore necessary to reactivate all investigations with that new product because it has a very different fatty acid composition (lack of long-chain ω-3 fatty acids such as EPA or DHA).

What are the recent results obtained with phosphatidylserine? A moderate improvement of short-term memory performances has been repeatedly observed in the elderly suffering with some cognitive deficits. A Japanese study of 78 subjects ranging from 50 to 69 years old and supplemented daily with 300 mg of soya phosphatidylserine showed a significant improvement in memory performances (Kato-Kataoka et al. 2010). However, it should be noted that phosphatidylserine did not generally bring any improvement in patients having an early degenerative dementia or Alzheimer's disease.

Experiments reported in 2010 by V. Vakhapova in Israel, using phosphatidylserine enriched in marine EPA and DHA (100 mg/day), showed after a 15-week treatment a significant improvement in short- and long-term memory in the elderly (Vakhapova et al. 2010). The improvement seemed more significant when the cognitive status of the persons was minimally affected before the treatment. Therefore, the outcome advocates the necessity of an early treatment of memory complaints.

An assay carried out in 2011 by A. G. Parker with 18 American students clearly showed an improvement of cognitive function (subtraction test) after a daily ingestion of 400 mg of soybean phosphatidylserine for 2 weeks (Parker et al. 2011). A similar experiment had been conducted in 2008 by J. Baumeister in Germany but this author was unable to detect any effect on cognitive functions. However, specific changes were observed using electroencephalography, indicating a new relaxation state after a period of stress. These effects may be helpful to people preparing for intellectual events requiring concentration.

A Japanese laboratory looked for the effect of a daily intake of 200 mg of soybean phosphatidylserine for 2 months in young children (average age 9 years) (Hirayama et al. 2013). The researchers demonstrated a beneficial effect only for short-term auditory memory, with the test consisting of measuring a repeated series of increasingly longer numbers. The result, although modest, should be considered important in a school setting for young children, with the effect of phosphatidylserine being able to improve, for example, the study of reading.

It is obvious that these experiments do not definitely allow the results to be associated to a specific phospholipid constituent (serine, fatty acids) or to the whole lipid molecule. The mechanism of the phosphatidylserine action remains unknown, despite numerous assumptions.

Given the published results, the US Food and Drug Administration recognized in 2003 that phosphatidylserine added to food could improve cognitive functions and slow their deterioration in the elderly. Despite few new studies since then, the European Commission decided in 2011 that soybean phosphatidylserine could be used in food for special medical purposes (Decision 2011/513/EU of 19 August 2011). A review of all the scientific work on the effects of phosphatidylserine on human brain concluded that phosphatidylserine is useful in improving various forms of memory, the ability to focus attention and concentrate, the ability to reason and solve problems, language skills, and the ability to communicate (Glade and Smith 2015). Although these memory effects are usually modest, they would benefit greatly from investigations over long periods (ca. 1 year) with chemically defined products, psychologically well-controlled patients, and several memory tests. On these issues, it would be important to define the effectiveness of a phosphatidylserine supplementation as a preventive or curative treatment. The abundance on the market of food supplements fortified with phosphatidylserine from various origins should motivate specialized laboratories to initiate serious behavioral and neuropsychological studies. Such work could clearly inform consumers about the potential benefits of such lipid supplementation marketable at a reasonable price.

3.3 Vitamin D

As for multiple sclerosis (Section 5.2.2), epilepsy (Section 5.3.2), schizophrenia (Section 6.1.3.2), and autism (Section 6.1.5.2), cognitive development during childhood seems to depend on the month of birth of the subjects, additional evidence of a possible relationship with the vitamin D production by skin under the effect of solar radiation (Section 2.2). Recall that on an academic basis, many studies have shown that children born in summer had more learning problems than children form in the other seasons, with these differences persisting until the age of 10–11 years (Martin et al. 2004).

A systematic review of 10 human studies has shown that only subtle cognitive impairments were observed in offspring of vitamin D–deficient mothers (Pet et al. 2016). However, validations of these findings are required.

Outside the context of pregnancy, very few studies have been devoted to the relationship between vitamin D and cognitive functions in children or adolescents. This is an unfortunate situation because no hypothesis may be proposed for possible improvements of such higher brain functions with that vitamin, a guarantee of better health in adulthood. The interest of such studies in young age is to avoid the interference from other factors such as the consumption of alcohol, drugs, or tobacco, behaviors known to impair cognitive performances.

The limited research performed with adolescents has failed to highlight an association between cognitive performances and vitamin D blood levels. One of the first investigations recruited almost 1700 teenagers as part of the great American survey on health and nutrition program (National Health and Nutrition Examination Survey), several psychometric tests, and a blood analysis being processed in each subject. Contrary to expectations, the results suggested that the vitamin D status was not related to cognitive performances in young subjects (McGrath et al. 2010). Similar outcomes were reported based on a study of a second sample of about 1800 teenagers as part of the same American survey (Tolppanen et al. 2010).

Besides these negative results, interestingly, an investigation was done on English teenagers comparing the educational outcomes and the serum levels of the two types of vitamin D, ergocalciferol (vitamin D_2) and cholecalciferol (vitamin D_3) (Tolppanen et al. 2012). Recall that vitamin D_2 comes from dietary sources (plants), whereas vitamin D_3 is mainly synthesized by the skin under sunlight exposure (Section 7.4). Tolppanen et al. (2012) showed that in contrast to vitamin D_3, the concentration of vitamin D_2 was inversely related to academic performances. Unfortunately, no hypothesis has been proposed to date to explain these results. A verification with further research could have important implications for vegetarians or vegans excluding any consumption of fish or egg, but depending only on a regular sun exposure to provide them vitamin D_3.

References

Bauer, I., Hughes, M., Rowsell, R., et al. 2014. Omega-3 supplementation improves cognition and modifies brain activation in young adults. *Hum. Psychopharmacol.* 29:133–44.

Baumgartner, J., Malan, L., Smuts, C.M. 2015. Novel interactions between iron and n-3 fatty acids in cognition and immune function. *Lipid Technol.* 27, No. 8. DOI: 10.1002/lite.201500042.

Baym, C.L., Khan, N.A., Monti, J.M., et al. 2014. Dietary lipids are differentially associated with hippocampal-dependent relational memory in prepubescent children. *Am. J. Clin. Nutr.* 99:1026–32.

de Groot, R.H.M., Hornstra, G., Jolles, J. 2007. Exploratory study into the relation between plasma phospholipid fatty acid status and cognitive performance. *Prostaglandins Leukot. Essent. Fatty acids* 76:165–72.

de Groot, R.H., Ouwehand, C., Jolles, J. 2012. Eating the right amount of fish: Inverted U-shape association between fish consumption and cognitive performance and academic achievement in Dutch adolescents. *Prostaglandins Leukot. Essent. Fatty Acids* 86:113–17.

Dyall, S.C. 2015. Long-chain omega-3 fatty acids and the braln: A review of the independent and shared effects of EPA, DPA and DHA. *Front. Aging Neurosci.* 7:52.

Eilander, A., Hundscheid, D.C., Osendarp, S.J., et al. 2007. Effects of n-3 long chain polyunsaturated fatty acid supplementation on visual and cognitive development throughout childhood: A review of human studies. *Prostaglandins Leukot. Essent. Fatty acids* 76:189–203.

Fontani, G., Corradeschi, F., Felici, A. et al. 2005. Cognitive and physiological effects of Omega-3 polyunsaturated fatty acid supplementation in healthy subjects. *Eur. J. Clin. Invest.* 35:691–9.

Glade, M.J., Smith, K. 2015. Phosphatidylserine and the human brain. *Nutrition* 31:781–6.

Hirayama, S., Terasawa, K., Rabeler, R., et al. 2013. The effect of phosphatidylserine administration on memory and symptoms of attention-deficit hyperactivity disorder: A randomised, double-blind, placebo-controlled clinical trial. *J. Hum. Nutr. Diet.* 27 (suppl. 2):284–91.

Holman, R.T., Johnson, S.B., Hatch, T.F. 1982. A case of human linolenic acid deficiency involving neurological abnormalities. *Am. J. Clin. Nutr.* 35:617–23.

Huang, T.L., Zandi, P.P., Tucker, K.L., et al. 2005. Benefits of fatty fish on dementia risk are stronger for those without APOE epsilon4. *Neurology* 65:1409–14.

Jackson, P.A., Forster, J.S., Bell, J.G., et al. 2016. DHA supplementation alone or in combination with other nutrients does not modulate cerebral hemodynamics or cognitive function in healthy older adults. *Nutrients* 8:86. DOI: 10.3390/nu8020086.

Jackson, P.A., Reay, J.L., Scholey, A.B., et al. 2012. Docosahexaenoic acid-rich fish oil modulates the cerebral hemodynamic response to cognitive tasks in healthy young adults. *Biol. Psychol.* 89:183–90.

Joffre, C., Nadjar, A., Lebbadi, M., et al. 2014. n-3 LCPUFA improves cognition: The young, the old and the sick. *Prostaglandins Leukot. Essent. Fatty acids* 91:1–20.

Johnston, D.T., Deuster, P.A., Harris, W.S., et al. 2013. Red blood cell omega-3 fatty acid levels and neurocognitive performance in deployed U.S. Service members. *Nutr. Neurosci.* 16:30–8.

Karr, J.E., Grindstaff, T.R., Alexander, J.E. 2012. Omega-3 polyunsaturated fatty acids and cognition in a college-aged population. *Exp. Clin. Psychopharmacol.* 20:236–42.

Kato-Kataoka, A., Sakai, M., Ebina, R., et al. 2010. Soybean-derived phosphatidylserine improves memory function of the elderly Japanese subjects with memory complaints. *J. Clin. Biochem. Nutr.* 47:246–55.

Kirby, A., Woodward, A., Jackson, S., et al. 2010. A double-blind, placebo-controlled study investigating the effects of omega-3 supplementation in children aged 8– 10 years from a mainstream school population. *Res. Dev. Disabil.* 31:718–30.

Konagai, C., Yanagimoto, K., Hayamizu, K., et al. 2013. Effects of krill oil containing n-3 polyunsaturated fatty acids in phospholipid form on human brain function: A randomized controlled trial in healthy elderly volunteers. *Clin. Interv. Aging* 8:1247–57.

Ladd, S.L., Sommer, S.A., LaBerge, S., et al. 1993. Effect of phosphatidylcholine on explicit memory. *Clin. Neuropharmacol.* 16:540–9.

Lassek, W.D., Gaullin, S.J. 2011. Sex differences in the relationship of dietary Fatty acids to cognitive measures in American children. *Front. Evol. Neurosci.* 3, article 5.

Maggioni, M., Picotti, G.B., Bondiolotti, G.P., et al. 1990. Effects of phosphatidylserine therapy in geriatric patients with depressive disorders. *Acta Psychiatr. Scand.* 81:265–70.

Martin, R.P., Foels, P., Clanton, et al. 2004. Season of birth is related to child retention rates, achievement, and rate of diagnosis of specific LD. *J. Learn. Disabil.* 37:307–17.

McGrath, J., Scragg, R., Chant, D., et al. 2010. No association between serum 25-hydroxyvitamin D3 level and performance on psychometric tests in NHANES III. *Neuroepidemiology* 29:49–54.

McNamara, R.K., Able, J., Jandacek, R., et al. 2010. Docosahexaenoic acid supplementation increases prefrontal cortex activation during sustained attention in healthy boys: A placebo-controlled, dose-ranging, functional magnetic resonance imaging study. *Am. J. Clin. Nutr.* 91:1060–7.

Montgomery, P., Burton, J.R., Sewell, R.P., et al. 2013. Low blood long chain omega-3 fatty acids in UK children are associated with poor cognitive performance and behavior: A cross-sectional analysis from the DOLAB study. *PLoS One* 8:e66697.

Muthayya, S., Eilander, A., Transler, C., et al. 2009. Effect of fortification with multiple micronutrients and n-3 fatty acids on growth and cognitive performance in Indian school children: The CHAMPION (Children's Health and Mental Performance Influenced by Optimal Nutrition) Study. *Am. J. Clin. Nutr.* 89:1766–75.

Nagata, T., Yaguchi, T., Nishizaki, T. 2011. DL- and PO-phosphatidylcholines as a promising learning and memory enhancer. *Lipids Health Dis.* 10:25.

Narendran, R., Frankle, W.G., Mason, N.S., et al. 2012. Improved working memory but no effect on striatal vesicular monoamine transporter type 2 after omega-3 polyunsaturated fatty acid supplementation. *PLoS One* 7:e 46832.

Parker, A.G., Gordon, J., Thornton, A., et al. 2011. The effects of IQPLUS Focus on cognitive function, mood and endocrine response before and following acute exercise. *J. Int. Soc. Sports Nutr.* 8:16.

Parletta, N., Cooper, P., Gent, D.N., et al. 2013. Effects of fish oil supplementation on learning and behaviour of children from Australian Indigenous remote community schools: A randomised controlled trial. *Prostaglandins Leukot. Essent. Fatty acids* 89:71–9.

Pet, M.A., Brouwer-Brolsma, E.M. 2016. The impact of maternal vitamin D status on offspring brain development and function: A systematic review. *Adv. Nutr.* 7:665–78.

Portillo-Reyes, V., Pérez-García, M., LoyaMéndez, Y., et al. 2014. Clinical significance of neuropsychological improvement after supplementation with omega-3 in 8– 12 years old malnourished Mexican children: A randomized, double-blind, placebo and treatment clinical trial. *Res. Dev. Disabil.* 35:861–70.21.

Richardson, A.J., Burton, J.R., Sewell, R.P., et al. 2012. Docosahexaenoic acid for reading, cognition and behavior in children aged 7–9 years: A randomized, controlled trial (the DOLAB Study). *PLoS One* 7:e43909.

Sheppard, K.W., Cheatham, C.L. 2013. Omega-6 to omega-3 fatty acid ratio and higher-order cognitive functions in 7- to 9-y-olds: A cross sectional study. *Am. J. Clin. Nutr.* 98:659–67.

Stonehouse, W., Conlon, C.A., Podd, J., et al. 2013. DHA supplementation improved both memory and reaction time in healthy young adults: A randomized controlled trial. *Am. J. Clin. Nutr.* 97:1134–43.

Tolppanen, A.M., Sayers, A., Fraser, W.D., et al. 2012. Association of serum 25-hydroxyvitamin D2 and D3 with academic performance in childhood: Findings from a prospective birth cohort. *J. Epidemiol. Community Health* 66:1137–42.

Tolppanen, A.M., Williams, D., Lawlor, D.A. 2010. The association of circulating 25-hydroxyvitamin D and calcium with cognitive performance in adolescents: Cross-sectional study using data from the third National Health and Nutrition Examination Survey. *Paediatr. Perinat. Epidemiol.* 25:67–74.

Vakhapova, V., Cohen, T., Richter, Y., et al. 2010. Phosphatidylserine containing omega-3 fatty acids may improve memory abilities in nondemented elderly with memory complaints: A double-blind placebo-controlled trial. *Dement. Geriatr. Cogn. Disord.* 29:467–74.

van der Wurff, I.S., von Schacky, C., Berge, K., et al. 2016a. Association between blood omega-3 index and cognition in typically developing Dutch adolescents. *Nutrients* 8:13.

van der Wurff, I.S., Bakker, E.C., Hornstra, G., et al. 2016b. Association between prenatal and current exposure to selected LCPUFAs and school performance at age 7. *Prost. Leukotr. Essent. Fatty Acids* 108:22–9.

Whalley, L.J., Deary, I.J., Starr, J.M., et al. 2008. n-3 Fatty acid erythrocyte membrane content, APOE varepsilon4, and cognitive variation: An observational follow-up study in late adulthood. *Am. J. Clin. Nutr.* 87:449–54.

Wild, K., Howieson, D., Webbe, F., et al. 2008. Status of computerized cognitive testing in aging: A systematic review. *Alzheimer's Dement.* 4:428–37.

chapter four

Cognitive decline and Alzheimer's disease

The cognitive decline that accompanies aging of all individuals is a normal and complex process, but slow and progressive, and probably a consequence of many factors. Some of these factors are known to facilitate the decline as hypertension, whereas others show it as a decline in physical activities or some hormonal treatments. In certain individuals, these disorders have a more rapid time course leading to very important cognitive disturbances, frequently marked by memory problems and sometimes accompanied by a worsening of executive functions. This stage is called mild cognitive impairment (MCI). Some of these problems will evolve into a state of mental disorganization, also known as dementia, with Alzheimer's disease being one of the main forms. The diagnosis of MCI is increasing in older people, and its incidence often follows simultaneously the increase in life expectancy. Thus, it is assumed that the incidence of this disease generally increases exponentially with age: 1 person in 10 at 65 years old and 1 person in 3 at 85 years old. This acute health problem requires the rapid identification of factors that could allow for control of its development, such as food constituents.

MCI, which affects a significant portion of the elderly population, has been explored to slow the decline progression, mainly via examining the influence of ω-3 fatty acids (Section 4.1.1), vitamin A, or carotenoids (Section 4.1.2).

Increasing research effort also has been devoted to the dementia syndrome of the Alzheimer's type. Indeed, this disease has been largely investigated in the field of ω-3 fatty acids (Section 4.2.1), vitamin A (Section 4.2.2), vitamin D (Section 4.2.3), vitamin E (Section 4.2.4), and cholesterol (Section 4.2.5).

4.1 Age-related decline

The first symptoms appear probably at the age of 45, as shown by a recent Franco–British study of the Whitehall II cohort (Singh-Manoux et al. 2012), and then become obvious at 65–70 years old. The decline clearly results from numerous structural and functional modifications in the brain. Alterations of the interactions between the subject and the activities of everyday life, and especially with the environment, may also be important.

It is now well established that aging is accompanied by a decrease of cognitive performance in most areas, with these decreases comparable to the decline observed in sensory functions. This decline of age-related cognitive capacities results from biological alterations of cerebral tissue. Indeed, the brain volume is progressively decreasing in parallel with important cellular changes, such as loss of dendrites and synapses. Modification of the cerebral microcirculation also is observed, leading to a decreased blood flow in the trophic capillary bed. Similarly, microvascular accidents and cerebral atherosclerosis have been found to induce neuronal loss in strategic areas.

Other less known factors were also discussed, such as an alteration of the transports through the blood-brain barrier or the disturbance of neurotransmitter systems, optionally combined with a demyelination process. It must be emphasized that these biological phenomena related to aging are not automatically accompanied by cognitive losses, many elderly having no cognitive impairment or they do not complain. Modern imaging techniques have helped explain these issues by brain plasticity, a recently known phenomenon that contributes to adapt the elderly to his environment.

The memory disorders observed in the elderly, who are considered otherwise healthy, are usually benign, simple memory lapses but whose frequency could increase with aging. These minor problems lead, in general, to complaints, but they do not necessarily correspond to an objective memory issue. It was known that in this field the age-related decline relates to working memory (particularly spatial memory), attention, and especially episodic memory (Ronnlund et al. 2005). Conversely, other forms of declarative memory, such as those concerning vocabulary and verbal IQ, seem to remain relatively intact in the elderly.

In a poorly defined part of the population, the loss of mental abilities during aging is more important than commonly found in most situations; we are dealing here with MCI. The proportion (prevalence) of affected individuals is currently high, between 3% and 22% annually according to the studies. This means that from 10% to 25% of individuals more than 70 years old in most countries are suffering from MCI. Overall, these troubles are constantly growing due to the aging of population, thereby stimulating the interest to better understand the pathological conditions involved. The MCI concept corresponds to a confounding zone between cognitive changes linked to normal aging and early stages of dementia. The distinction between normal aging and MCI remains difficult to establish and requires the use of clinically valid tools. In addition, there is controversy about the need to distinguish MCI subgroups based on the number of affected cognitive domains. Concrete progress is therefore expected to better predict the evolution of these disorders and define the future experimental protocols.

That step in the evolution of disorders is marked by a worsening memory complaint from the patient and from family. Although difficult to characterize, these troubles have anatomical features in common with the state of dementia, such as the presence of amyloid deposits and a volume reduction of the hippocampus. At this stage of the disease, these anatomical lesions are rarely revealed by the use of medical imaging techniques (especially MRI). In current practice, the doctor verifies whether the patient's memory complaints (or those around him or her) are indeed associated with real memory disorders that could be later detected with a clinical examination based on more or less specialized tests such as the Mini-Mental State Examination (MMSE) or the Trail Making Test (TMT) (Section 7.9). These assessment tools are selected to estimate the global intellectual efficiency and help the clinicians monitor the development of the disease.

The presence of MCI, affecting mainly the episodic memory (memory of autobiographical events), and sometimes also the executive memory, actually corresponds to an early stage of nervous degeneration, but it does not consistently move toward a state of dementia. Thus, it was observed that approximately 7%–12% of the subjects by year displaying these troubles will progress toward Alzheimer's disease. Four years after making the diagnosis of MCI, approximately 50% of the patients will be suffering from dementia.

As for Alzheimer's disease, the MCI condition is characterized by significant and irreversible brain alterations in defined areas.

No current therapy exists to fight against MCI. However, this disease is considered as the optimal stage for interventional studies aimed at delaying the onset of dementia. Given the importance of lipids in the structure and functions of the brain, it seems urgent to consider nutrition as an opportunity to slow the progression of the age-related cognitive disorders to reduce or even offset the risk of developing dementia. Among nutrients, several lipid compounds have already been the target of investigations: ω-3 fatty acids (Section 4.1.1) and vitamin A and carotenoids (Section 4.1.2).

4.1.1 ω-3 Fatty acids

Scientists began in the early 1980s to study the role of nutrition in cognitive processes. But it was not until the late 1990s that essential advances were made with the so-called "Rotterdam" study. That important survey study demonstrated that the consumption of ω-3 fatty acid–rich fish enhanced cognitive performances in the elderly, whereas an excess linoleic acid (precursor of ω-6 fatty acids) had the opposite effect (Kalmijn et al. 1997). This thesis, although still disputed, has since been verified many times by independent research teams fighting against dementia and Alzheimer's disease. Several researchers have also discussed the possible action of ω-3 fatty acids on the brain vascular system, as it has long been demonstrated for the cardiovascular system (Leray 2015).

In the elderly, what are the possible relationships between ω-3 fatty acids and cognitive functions? Could nutrients delay the alteration of these functions or affect that development throughout life in maintaining nerve structures? What is known about the effects of the supplementation with docosahexaenoic acid (DHA) combined or not with eicosapentaenoic acid (EPA)? Clinicians are interested in the population aged over 50 years, because complaints of memory loss are becoming more and more frequent.

4.1.1.1 *Epidemiological investigations*

Since the beginning of the twenty-first century, numerous studies have shown that fish consumption would help to prevent the age-associated cognitive decline and perhaps even the dementia that is often associated with it. Meanwhile, it has been verified that the ω-3 fatty acid content in brain tissue decreased with advancing age. This poorly explained phenomenon could be not only connected to a reduction in the transport of these fatty acids from the blood to the brain but also to an alteration of their metabolism in the brain tissue itself. It can also be hypothesized that the elderly, living at home or in nursing institutions, progressively decrease their seafood consumption or have less and less efficient intestinal absorptive capacity. These changes could thus contribute to reducing the DHA brain concentration. It has been demonstrated that older animals have a specific reduced DHA concentration in the hippocampus, the brain location known to play a key role in memory and spatial orientation. Thus, the links between the chemical composition of nervous tissue and memory capacities are becoming clearer.

In the fight against the cognitive decline associated with age, the benefits brought about by fish consumption are difficult to estimate, although a trend is gradually emerging. Some work concerning MCI details the benefits of frequent or continuous consumption of marine fish that could help adults remain healthy. MCI is a syndrome defined by clinical, cognitive, and functional criteria that implies, according to specialists, a predementia condition (Section 7.9). This syndrome affects almost exclusively the memory and was found to be a good predictor of a progression to dementia (10%–15% of cases) during the year after diagnosis. The clinical diagnosis of dementia is retained only when the cognitive or behavioral disorders seriously disrupt the patient's daily life. Some epidemiological and neuropsychological studies in many patients have shown that fish consumption was associated with a reduced risk of cognitive function loss (memory, comprehension, psychomotor speed). Of course, as very often in clinical research, other studies did not fully confirm this association, but fortunately detected no particular negative effect.

One of the studies showing the most favorable effects of a fish-enriched diet on the cognitive capacities of the elderly is that achieved in Norway in 2007 through a collaboration between Bergen and Oslo universities (Nurk et al. 2007). An accurate survey of dietary habits and examinations with appropriate psychometric tests (including MMSE and TMT, Section 7.9) were performed on nearly 2000 individuals aged from 70 to 74 years. The results showed that the subjects consuming an average of more than 10 g per day of fish or seafood had significantly higher scores in six tests, thus they had better cognitive performance than subjects consuming lower amounts. This daily 10-g threshold corresponds in reality to only one fish meal per week. The results also showed that maximum effects are associated with a consumption of about 75 g of fish per day. Conversely, other studies did detect some positive effects for a more limited number of neuropsychological tests.

More recently, four investigations have been done in France, the USA, and China:

- The French study, published in 2009, is part of a wide epidemiological investigation on women registered in a mutual life insurance company (Mutuelle Générale de l'Education Nationale [MGEN]) (Vercambre et al. 2009). Approximately 4800 women aged 76–82 years filled out a self-administered questionnaire and underwent neuropsychological tests. Outcomes clearly supported the hypothesis of a beneficial effect of foods rich in ω-3 fatty acids in the prevention of the cognitive decline observed during aging. That outcome was confirmed by another interesting multidisciplinary study of about 3300 subjects of both sexes and 64 years average age. They were selected from the French study SUVIMAX (Supplementation with Vitamins, Minerals and Antioxidants) (Kesse-Guyot et al. 2011).
- The American study, published in 2013, recruited 6000 women, with mean age of 72 years who worked in the health field (Women's Health Study) (Kim et al. 2013). The authors observed better cognitive capacities, mainly in verbal memory, in people consuming one or more fish servings weekly. The fish in question were dark-meat fish (e.g., tuna, mackerel, sardines, salmon, or swordfish). All of these species are particularly rich in ω-3 fatty acids. Conversely, no benefit was observed with light-meat fish (e.g., cod, haddock, or halibut) as well as shellfish (e.g., lobster, scallop, or shrimp), all of which are very poor in ω-3 fatty acids.

- The Chinese study, published in 2014, was realized by two American teams that specialized in human nutrition and psychiatry (Qin et al. 2014). The results were obtained from approximately 1600 community-dwelling adults older than 55 years, with a mean follow-up of 5.3 years in nine Chinese provinces. They observed that a weekly consumption of more than one fish meal (about 100 g) in subjects older than 65 years reduced the mean annual rate of global cognitive decline by 0.35 point, a gain equivalent to the disparity associated with 18 months of age.

An original approach also has been carried out by a research team in Uppsala, Sweden, that established a positive correlation between EPA and DHA dietary intake, overall cognitive performances, and the gray matter volume measured by MRI (Titova et al. 2013). The objectivity of the imaging measurements in the brain demonstrated at least an unquestionable anatomical effect, even if the results of exploration are still the subject to discussions. These results were confirmed 1 year later with the same techniques by clinicians at the University of Pittsburgh, Pennsylvania (Raji et al. 2014). This work showed with more fish consumption, there was a greater mass of gray matter (neurons) in areas responsible for cognition. Actually, nothing can proved regarding the direct influence of ω-3 fatty acids on these anatomical variations. Apart from differences in lifestyle among fish eaters, no serious hypothesis can now explain these results. Very similar results on the volume of the whole brain and the hippocampus together with the average cortex thickness were recently published by a team from Columbia University, New York (Gu et al. 2015). The authors calculated that in the elderly, a weekly consumption of 90–150 g of fish would provide significant protection against brain atrophy, a benefit equivalent to 3–4 years of aging. Thus, an effortless dietary modification may ensure a significant effect of maintaining cognitive functions.

This subject remains highly topical and of such importance that two major analyses were published in early 2015. One analysis of more than 20 studies done over about 15 years in this field allowed the conclusion that ω-3 fatty acids provided by fish have a more or less important beneficial effect on cognitive activity during aging (Denis et al. 2015). The other analysis of 15 studies selected among 1034 publications revealed that adults taking at least 1 g/day of an EPA plus DHA mixture improved episodic memory, with DHA being specifically effective for semantic memory (Yurko-Mauro et al. 2015). These results are important because they help us to understand why many investigators have detected no effect with smaller fatty acid amounts. Notably, this threshold of 1g/day is not excessive; it corresponds to a daily consumption of 50 g of salmon or 100 g of sardines.

As for many other physiological aspects of dietary lipids, it is necessary to consider the relative amounts of the ω-6 fatty acid intake in addition to that of ω-3 fatty acids (Loef and Walach 2013). Indeed, many studies have shown that the absolute amount of ω-3 fatty acids is a less important marker than the ω-6-to-ω-3 fatty acid ratio, an index measured in diet or in blood. This concept is not new for nutritionists (Leray 2015), but it has only recently been applied to the relationships between dietary lipids and cognitive decline or dementia. Among the first work raising this issue is the French study "Trois-Cités" (Bordeaux, Dijon, Montpellier). The study enrolled around 10,000 subjects and showed that a high consumption of ω-6 fatty acid–rich oils was associated with a higher risk of dementia and even Alzheimer's disease. Notably, these adverse effects are all the more marked in that the intake of ω-3 fatty acids is low, and even more pronounced among the apolipoprotein E4 (ApoE4) noncarriers (Barberger-Gateau et al. 2007).

Given the well-known metabolic competition between these two fatty acid series, it is not surprising that in human history the change from a prehistoric diet (with a ω-6/ω-3 fatty acid ratio close to 1) to a modern diet, or "Western" diet (with a ω-6/ω-3 fatty acid ratio sometimes close to 30), may influence brain function. Although it is not currently possible to make specific recommendations, it seems more desirable to first decrease the intake of vegetable sources providing too much ω-6 fatty acids before increasing the intake of ω-3 fatty acids.

Unfortunately, epidemiological studies have usually resulted shown great variability of results, likely arising from the poor reliability of the information provided by consumer food surveys. The estimation errors were often placed on the account of the failing memory of surveyed subjects, the imprecision of food composition tables, and the culinary practices being different between countries and people. The protective effects of DHA and EPA may be also insidiously underestimated due to an increased mortality from cardiovascular disease in people consuming only small amounts of fish. This indirect consequence of a diet deficient in ω-3 fatty acids could thus reduce the chances of observing cognitive issues in the elderly, who somehow died prematurely.

These approximations and some others may be the cause of the finding of an absence of relationships between the oily fish consumption and cognitive performances among US veterans, despite a 3-year follow-up and the use of a large number of psychometric tests (van de Rest et al. 2009).

Regardless, these epidemiological studies cannot eliminate the possibility of an association between fish consumption and socioeconomic level, which led to select subjects accustomed to a balanced diet and having a lifestyle without any excess. These factors are well known to minimize the risk of cognitive impairment.

To complete this aspect of research, the detrimental consequences of pollution by fish mercury must also be considered. Indeed, such pollution has recently been evoked to explain some adverse effects of a high fish consumption on the cognitive performances in the elderly living in Adelaide, South Australia, an area known for its high pollution rate (Danthiir et al. 2014). However, this risk may be minimized as a recent study conducted at the Rush Medical University of Chicago, United States, clearly showed that for a moderate consumption of fish (up to three servings weekly), the brain mercury levels were not correlated with any neuropathological marker (Morris et al. 2016). Similarly, mercury exposure was found to have little impact on cognitive performance in elderly men and women from Kuopio, Finland (D'Ascoli et al. 2016).

As shown by the ω-3 fatty acid analysis in plasma or red blood cells, it becomes possible to reduce the sources of error related to dietary questionnaires. Thus, a study of 246 participants aged 63–74 years living in Nantes, France, showed after a 4-year follow-up that a higher proportion of DHA (as opposed to EPA) in erythrocyte membranes was significantly associated with a lower cognitive decline (Heude et al. 2003). In addition, a similar study conducted in the United Kingdom confirmed that high levels of ω-3 fatty acids in erythrocytes were closely correlated with a slowdown of cognitive decline, but only in individuals not carrying the ApoE4 gene (Whalley et al. 2008). Comparable results were published from a recent Japanese study of 400 subjects aged 60–79 years and followed up for 10 years (Otsuka et al. 2014). Conversely, the finding of a similar relationship, but only in relation to the ApoE4 gene carriers, may result from not taking into account the dietary supply of ω-6 fatty acids (Morris et al. 2016). That explanation may also be valid for the observation of slower rates of decline in global cognition with weekly seafood consumption in elderly of average age 81 years (van de Rest et al. 2016).

An interesting aspect is that addressed by the Brain Institute at the University of Oregon in Portland (Bowman et al. 2013). The team of Dr. J. F. Quinn showed that in older people (about 74 years old) followed up for 4 years, the executive function decline (estimated by the TMT, Section 7.9) was associated with a low concentration of blood ω-3 fatty acids and a higher volume of subcortical white matter. That volume increase has long been considered as a marker of cognitive decline and motor dysfunction. They were thus able to estimate that for each 100 μg/mL increase of plasma ω-3 fatty acids, the age-related cognitive decline is delayed by 1 year, without alteration of verbal memory or MMSE test scores.

Recent work has shown that in elderly people without mental disability, a high blood level of ω-3 fatty acids was accompanied by a higher cognitive flexibility and a greater gray matter volume in the anterior cingulate area, known to be important for that function (Zamroziewicz et al. 2015).

A more recent study was done in Finland with 768 participants with a mean age of 68 years, but without cognitive impairment, who were examined at baseline for their cognitive function and serum fatty acid composition and then reexamined 11 years later (D'Ascoli et al. 2016). The authors have found that people with higher serum EPA plus DHA, reflecting mainly fish consumption, had better performance in the TMT (Section 7.9) and the Verbal Fluency Test, with both tests being specific for frontal lobe functioning. Notably, the associations with DHA were stronger and the ApoE4 phenotype had little impact on cognitive performance. These studies are the first involving the precision of ω-3 fatty acids as specific factors linking nutrition and cognitive functions.

Many other studies have been devoted to this problem, but the variability in results and the diversity of the psychometric tests used require caution in interpreting the findings and announcing recommendations. The importance of the dietary ω-6 fatty acid supply in the scale and diversity of the outcomes remains a frequently topic of discussion. Despite these reservations, it seems that the blood ω-3 fatty acid composition is a reliable biochemical marker of the cognitive impairment related to aging, even in the absence of dementia or cardiovascular disease (Tan et al. 2012).

It is evident that the meaning of the blood DHA status remains unclear: is it a reflection of the nutritional intake or that of the metabolism, or a better intestinal absorption, or a lower cellular catabolism? These questions are fundamental for studies related to the elderly. Future research in this area should take into account these physiological aspects while comparing them with the genetic characteristics of the selected participants.

Despite the reservations expressed by some authors and the lack of consensus on the positive effects of ω-3 fatty acids, it seems wise to advise the entire population to increase the presence of fish meals on the family table so that parents and children can benefit from a sufficient ω-3 fatty acid contribution. By simply following the recommendations of the French National Plan of Nutrition and Health, consumers will obtain substantial health benefits for cognitive aging (PNNS 2011-2015 report, available at http://www.sante.gouv.fr/programme-nationalnutrition-sante-2011-2015.html).

The US Food and Drug Administration and the US Environmental Protection Agency recently issued updated advice on fish consumption. The two agencies have concluded that pregnant and breastfeeding women, those who might become pregnant, and young children should eat more fish with low mercury content to gain important developmental and health benefits. The updated advice recommends pregnant women eat at least 224 g (8 ounces) and up to 336 g (12 ounces) per week (2–3 servings) of a variety of fish that is known to be lower in mercury to support fetal growth and development. These weakly polluted species include some of the most commonly eaten fish, such as pollock, salmon, canned light tuna, tilapia, catfish, cod, and even shrimp (http://www.ncbi.nlm.nih.gov/books/NBK305180/).

Despite the presence of contaminants, the belief of many experts has been that consuming fish is beneficial for health. The "Dietary Guidelines for Americans, 2010" also recommends a consumption of 227 g (ca. 8 ounces) of seafood per week.

4.1.1.1.1 Intervention studies Given the existing problems limiting fish consumption (e.g., availability, pollution, taste), investigators have experimented with alternative sources such as fish oils or purified ω-3 fatty acids. So, what have we learned from that approach, the aim of which is to reduce efficiently the cognitive decline in aging?

Admittedly, intervention studies are not sufficient for clarifying the subject, likely because of the wide variety of selected protocols, the difficulty in evaluating neuropsychiatric symptoms, and their evolution during a relatively short time. A recent meta-analysis of three clinical trials by the international Cochrane Center pointed out that, given the results published from 2008 to 2011, a ω-3 fatty acid supplementation provides little profit for the cognitive function of the elderly in good mental health, as indeed for those suffering from dementia (Dangour et al. 2012).

Recently, thanks to the exceptional advances in human genetics, biologists better understand the complex mechanisms involved in how our cells absorb nutrients and react to food. Nutrigenomics is the result of that new alliance between genomics and nutritional sciences. This approach allows a better understanding of how nutrients influence the expression of our genome and also how the genome itself can influence the metabolism of these nutrients.

Thus, concerning the ω-3 fatty acids, it has been shown that the presence in the genetic heritage of an individual of the allele encoding the protein apolipoprotein epsilon 4 (or ApoE4 isoform) was closely associated to a high risk of cognitive decline and especially of Alzheimer's disease. As mentioned by A. M. Minihane in England, the ApoE4 gene could indeed interact with ω-3 fatty acids and promote the genesis of cardiovascular diseases and secondarily of brain function disorders (Minihane et al. 2000). The potential impact of this gene in the status of the cognitive performances in the elderly has been mentioned previously. Thanks to a large number of results, there is no longer any doubt that genes interact with the environment regarding cognitive functions (Small et al. 2004).

The validity of this explanation is based on the observation of a close association between the level of ω-3 fatty acids in red blood cells and the preservation of the cognitive performances in patients not carrying the ApoE4 gene (Whalley et al. 2008). Interestingly, it has been proved that only the individuals with mild cognitive decline, but lacking the ApoE4 gene, might benefit from fish oil ingestion to improve their cognitive status (Daiello et al. 2015).

Nevertheless, the mechanisms involved in the interactions among genes, foods, and cognitive performances are still poorly understood. Moreover, the influence of compounds other than lipids should not be ruled out. Future research should consider not only the nature and the amount of the ingested fatty acids but also the genetic status of ApoE in the patients examined. If these approaches are confirmed, any attempt to prevent a loss of the age-related cognitive performances should be preceded by a genetic study to increase the prevention efficiency. Actually, clinicians undertake these analyses only in the context of research on the treatment of a declared Alzheimer's disease, but it is likely that work carried out with ω-3 fatty acids will encourage health authorities to expand in the near future the indication of such investigations.

The variability of results reported in the scientific press is likely the cause of the heterogeneity of the participants of clinical tests. Thus, a British study in 2010 did not detect any change in cognitive performances in 800 subjects with a mean age of 75 years after a daily intake of 500 mg of DHA and 200 mg of EPA during 2 years (Dangour et al. 2010). Similarly in 2015, Dr. E. Y. Chew, as part of the Age-Related Eye Disease Study 2 (AREDS2), could not detect any beneficial effect after prescribing 350 mg of DHA and 650 mg of EPA for 5 years to a large number of individuals about 73 years old. These patients were selected for a general study of retinal disorders (Chew et al. 2015).

In fact, Dr. Chew noted that the dosage could be one of the factors that affected their findings. Importantly, the vast majority of studies using less than 700 mg of DHA daily have concluded no benefit, whereas studies using higher amounts indicated beneficial effects on cognitive function in adults ranging from healthy individuals to those having mild Alzheimer's disease (http://www.goedomega3.com/index.php/blog/2015/08/omega-3s-and-cognition-dosage-matters). However, the cause of the negative outcomes is probably multiple: advanced age of subjects, adequate dietary ω-3 fatty acid intake, or inadequate psychometric tests.

As for other experiments, the large number of enrolled patients does not always provide greater experimental rigor.

A remarkable investigation in this area was conducted in 19 US sites involving 485 healthy subjects, aged more than 55 years, with half being supplemented daily for about 6 months with 900 mg of DHA, and the others receiving a placebo (Yurko-Mauro et al. 2010). This large clinical trial, conducted with the help of Martek Biosciences Corporation that produced a seaweed DHA extract, has shown a remarkable beneficial effect on learning and episodic memory (verbal recognition), but not on spatial memory or pattern recognition. In addition, the cognitive test scores were significantly correlated with the blood DHA concentrations. The authors of this extensive work considered that the improvement observed after 6 months in individuals receiving DHA corresponded to the state observed in 3.5-year younger subjects.

Other work carried out on a smaller group of individuals receiving DHA daily for 3 months (Rondanelli et al. 2012) or on a larger group receiving fish oil for 1 year (Lee et al. 2013) has demonstrated the usefulness of such treatments in improving cognitive performances in the elderly suffering from MCI.

A rigorous meta-analysis of eight investigations in the same field done between 2006 and 2010 came to the same conclusions: the supply of ω-3 fatty acids is capable of inducing beneficial effects on the memory characteristics in healthy subjects having only memory complaints or scores indicating only a mild cognitive decline (Mazereeuw et al. 2012).

The subject remains highly topical and of such an importance that a new meta-analysis was published in early 2015. This analysis, encompassing 15 investigations selected from 1034 publications, found a real improvement in the episodic memory of adults taking at least 1 g/day of an EPA plus DHA mixture, with the DHA being particularly efficient for semantic and working memory (Yurko-Mauro et al. 2015).

To conclude the survey of this field of investigation, it is also necessary to take into account the results of rigorously designed clinical research recently performed in Germany at the Leibniz University of Hanover (Witte et al. 2014). The authors clearly showed that a treatment with ω-3 fatty acids from fish oil (2.2 g/day for 6 months) led to positive effects in brain function, mainly on its executive functions (TMT, Section 7.9) in the elderly without neuropsychological disorders (score MMSE >26). The treatment also showed a beneficial effect on the blood levels of neurotrophic factors and on the nerve microstructure appreciated by MRI techniques.

As was previously described for young adults (Section 3.1), new studies demonstrate that 24 weeks of fish oil supplementation (2.4 g/day of EPA plus DHA) increases prefrontal cortical hemoglobin oxygenation (blood oxygenation level–dependent signal) in healthy older adults (62–80 years), together with their working memory performance (Boespflug et al. 2016). These findings suggest that increasing dietary long-chain ω-3 fatty acids intake may help to reduce neurocognitive decline and more importantly neurodegenerative processes when initiated early in their onset.

Moreover, it has been proven that after a treatment with fish oil, unlike patients with Alzheimer's disease, only patients with a light cognitive decline, but lacking ApoE4, displayed improvement in their cognitive condition (Daiello et al. 2015). These results are in agreement with the conclusions reported by J. F. Quinn that people suffering from Alzheimer's disease (MMSE score between 14 and 26) and treated with 2 g of DHA/day did not see a slowdown of their cognitive decline (Quinn et al. 2010).

Despite certain limitations, all the research cited above suggests that a dose ranging from 1 to 2 g of DHA per day can slow cognitive decline in healthy subjects and also in those with mild forms of dementia. The mechanisms involved are poorly known, but they seem to aim specifically at the modulation of inflammation and apoptosis. This hypothesis is supported by a study proving that DHA is more effective than EPA in modulating specific markers of inflammation in humans (Allaire et al. 2016).

It is unfortunate that no treatment with purified EPA has been initiated. Although the state of research in this field does not allow targeting of particular populations according to their genetic or physiological predispositions, it is essential to maintain from a very young age an ω-3 fatty acid intake consistent with the most recent recommendations (Section 7.2). As emphasized by Dr. P. Barberger-Gateau at the University of Bordeaux in France, it is certain that upcoming experimental research will consider more parameters to target better defined populations on the point of view of cognition, nutrition, and genetics to optimize the future prescriptions (Barberger Gateau et al. 2013).

Five clinical projects are about to be completed or were still recruiting in 2016 in various research centers targeting the effects of ω-3 fatty acid supplementation on age-related cognitive decline (http://clinicaltrials.gov).

In conclusion, all this research generates new ways or means to prevent or delay the devastating effects of time on our cognitive performances. All authors invested in this fight emphasized the need to set up an early prevention, based on a regular consumption of fish and, if necessary, a daily intake of at least 1 g of purified ω-3 fatty acids. Any fatty acid supplementation should be applied well before the onset of dementia symptoms; it should be recommended from the age of 40 or 50 years for people at risk, knowing that clear criteria of the disease can be established only after the progression of the underlying disease remaining silent during several years.

4.1.2 Vitamin A and carotenoids

4.1.2.1 Vitamin A

Laboratory experiments over time have shown that retinoic acid, one of the forms of vitamin A (Section 7.3), plays a key role in the developing nervous system. In the Laboratory of Cognitive Neuroscience at the University of Bordeaux, Dr. F. Mingaud has shown, in animals, that vitamin A is involved in the regulation of synaptic plasticity (Mingaud et al. 2008). This plasticity is now taken into account to explain the modifications in learning and memory functions mainly by means of the expression of target genes. Other experiments conducted at the same university in young rats have shown that a dietary vitamin A supplementation improved the neuronal dendritic arborization and especially the spatial memory in older animals (Touyarot et al. 2013).

Despite these encouraging results, the actual role of retinoids in the human cognitive functions remains enigmatic, although the intervention of these lipid compounds in the emergence of dementia is now an undisputed fact (Section 4.2.2).

4.1.2.2 Carotenoids

Lutein, a widespread carotenoid in plants (Section 7.3) seems to be a compound potentially very important for cognitive functions. As zeaxanthin, it is known to cross readily the blood–brain barrier and accumulate in all nerve tissues, and specifically in the region of retina called the macula. Thus, the measurement of macular pigment concentration by photometric methods has been proposed to replace in blood the complex determination of these pigments.

The academic interest for that molecule is new; it originated more than 10 years ago during a US epidemiological study (Kang et al. 2005). This large trial (Nurses Health Study), involving more than 13,000 women, revealed that subjects consuming the greatest amount of green vegetables (lutein rich) had a cognitive decline less pronounced that subjects consuming green vegetables only rarely. The authors were able to estimate the gain in terms of "cognitive age" equivalent to about 2 years. More recently, the laboratory of Dr. Elizabeth J. Johnson, a recognized specialist in this issue at Tufts University in Boston, studied lutein as well as zeaxanthin and β-carotene in an important number of seniors and even in centenarians living in the US state of Georgia (Johnson et al. 2013). These investigations showed that the cognitive performances of all subjects were directly correlated with the blood levels of these three carotenoids, with lutein characterizing more particularly octogenarians. Similar results were obtained after analysis of brain samples taken from people who died during the study. These results have been repeatedly confirmed in many elderly after research based on the oral intake of β-carotene (Dutch study) or on the plasma concentration as well (Swiss study).

It is even more interesting to learn that in people older than 50 years the lowest macular pigment concentrations are correlated with the lower cognitive performances (Feeney et al. 2013). It is therefore possible to say, show me your eyes, and I shall tell you how your brain works.

Dr. E. J. Johnson published the results of a supplementation trial of 10 women, aged about 67 years and who were in good mental health, with lutein at 12 mg/day for 4 months. At the end of that study, the subjects receiving lutein showed a significant improvement in verbal fluency scores, unlike subjects receiving a placebo (Johnson et al. 2008). The test used allowed the assessment of long-term memory of subjects by asking them to list the greatest number of names belonging to the same category (flowers, animals, etc.) for 1 minute. Although these results were obtained with only a small number of individuals, they seemed encouraging to initiate further research on the therapeutic use of lutein, perhaps synergistically with other treatments such as ω-3 fatty acids.

However, an experimental study performed by Dr. E. Y. Chew, within the general American AREDS2 program, could not highlight any benefit after 5 years with a daily intake of 10 mg of lutein and 2 mg of zeaxanthin in people of average age 73 suffering from mild retinal disorders (Chew et al. 2015).

No specific carotenoid intake has been recommended in humans, but the few data collected to date on the effects of lutein, and perhaps zeaxanthin, on various forms of memory may only encourage all of us to consume, as much as possible, green vegetables such as kale and spinach, which are particularly rich sources of these compounds (Leray 2015). Recall that French people ingest with their food, on average, 2.5 mg/day lutein and zeaxanthin (Section 7.3), whereas 6 mg was associated with a reduced risk of age-related macular degeneration (AMD). Similar low dietary intakes of these carotenoids have been determined in Americans from Indianapolis, Indiana (Curran-Celentano et al. 2001). Efforts remain to be done to achieve the threshold dose that could be definitely beneficial to many functions, both in the visual field and the cognitive field.

4.2 *Dementia and Alzheimer's disease*

MCI progresses to dementia frequently. The clinical situation is then characterized by significant cognitive deficits, severe enough to affect family, social, or professional life. Thus, a progressive loss of memory is recorded, but also an alteration in motor movements and language and difficulties with thinking or problem-solving. The disease has been associated specifically with a loss of executive functions that characterize the execution of tasks requiring a goal or defined objectives. These high-level functions are involved in many forms of cognitive activities (Section 7.9).

Among dementias, a distinction should be made between Alzheimer's disease, frontotemporal dementia (Pick's disease), vascular dementia, and other dementia associated with certain diseases (Creutzfeld–Jacob, Parkinson, Huntington). Certain neurologists even argue that the majority of dementia in the elderly would be a trouble combining degenerative disease and vascular pathology. The term "senile dementia" was used when it was thought that memory loss and confusion were a normal part of aging. It is more common now to refer to dementia, or early-onset dementia, if the person is under 65 years old.

As Prof. Roger Gil (2010) said: "It is difficult to have an unitary approach of a set as disparate as organic dementia. This disparity is due to the symptomatology of dementia syndrome that is not one but is plural;

it also lies in the diversity of all etiologies as well as to the heterogeneity of clinical presentations within a same etiology."

Thus, in the absence of reliable biomarkers, clinicians must unfortunately wait for the full establishment of the disease before characterizing with success an advanced dementia, a time that can delay any intervention. Following the works of Prof. Bruno Dubois, Pierre and Marie Curie University in Paris, the current trend is to clinically characterize an Alzheimer's prodromal phase after a preclinical phase, a period still currently poorly characterized that lasts from 15 to 20 years. The prodromal stage shows symptoms including impairment of episodic memory of the hippocampal type that does not yet alter daily life and the presence of biomarkers detected in the cerebrospinal fluid (β-amyloid peptide) or by MRI (hippocampal atrophy, senile plaques).

The "World Alzheimer Report" estimated that in 2010 the number of people affected in the world by dementia was 35.6 million and will reach 66 million in 2030 and more than 115 million by 2050 if no medical treatment reduces the incidence of this disease. Globally, the costs incurred by dementia have been estimated at more than US$600 billion dollars.

The US Alzheimer's Association has estimated that more than 5 million Americans are living with Alzheimer's disease, with approximately 200,000 individuals being under 65 years. Notably, one in three seniors dies with Alzheimer's or another dementia. In 2015, the cost of the unpaid care for patients has an estimated economic value of more than US$220 billion. As emphasized by the association, Alzheimer's takes a devastating toll, not just on those with the disease but also on entire families. These numbers will increase rapidly in coming years, because the baby boom generation has begun to reach age 65 and beyond, the age range of greatest risk for the disease.

In Europe, all dementias affect about 6.4% of the population, with this figure being 1.2% between 65 and 69 years and nearly 30% after 90 years.

In France, there are currently about 900,000 patients, with an incidence of more than 100,000 new cases per year. The prospects for the coming years are quite dark because, considering the current trends, an INSERM epidemiologist team has estimated a 75% increase in dementia cases in the French population between 2010 and 2030 and 200% in people more than 90 years (Institut National de la Santé et de la Recherche Médicale [INSERM] Unit 897/Université Victor Segalen, Bordeaux 2). These alarming estimates are nevertheless necessary to anticipate the needs and manage this relentless evolution in our society. Ignoring the financial burden for families, the annual fee for the French patients is currently close to €19 billion (0.6% of gross domestic product). Despite these alarming figures, it has been estimated that if it were possible to delay the onset of the disease a year, more than 9 million people would avoid the disease in the world in 2050 (Wimo and Prince 2010).

The diagnosis of Alzheimer's disease is currently unclear and long to establish, because it is based solely on the use of neuropsychological tests to evaluate the patient's cognitive functions, and at different times of the evolution of the pathology. Frontotemporal dementia could represent about 20% of the degenerative dementia (about one third of the cases diagnosed in people under 65 years old) and 30% of patients would in fact suffer from vascular dementia, especially for the early cases. Recall that the increasingly frequent MRI scans show that 30% of patients older than 65 years have signs of cerebral infarction. The imprecision of diagnosis, due to the overlap of symptoms, makes the estimation of the contribution of each type of disease to the cognitive decline difficult, likely contributing to the dispersion of the results registered during epidemiological or clinical work.

In neuropsychology, Alzheimer's disease differs from mild cognitive decline by a movement toward a loss of autonomy followed by disability. In both cases, the first signs are identical, and the same psychometric tests, such as the MMSE and the TMT (Section 7.9), are often used to establish a diagnosis. Some patients have a rapid decline (loss of at least 3 MMSE points in 1 year) that has a poor prognosis, whereas others have a slower decline. The diagnosis of the disease may be established not only on the basis of an evolution of cognitive functions, by using, for example, the specific Alzheimer Disease Assessment Scale-cognitive (ADAS-cog, Section 7.9), but also more recently through imaging techniques (computerized tomography, MRI). These new procedures allow to appreciate a possible atrophy of specific brain areas, such as the hippocampus, but they will be soon complemented by positron emission tomography scans enabling the quantification of amyloid deposits and thus a more accurate diagnosis. Analysis of blood and cerebrospinal fluid may also be used. Research is nevertheless required to establish the validity of these promising new tools.

After the appearance of a mild cognitive decline, some patients present a progressive worsening of their cognitive functions, such as memory, judgment, decision-making, language, and orientation. Added to that list several must be behavioral issues such as anger, aggression, greed, distrust, depression, delirium, or hyperphagia. The neuropathologists add to the clinical picture many neuronal plasticity alterations, such as a selective loss of neurons and synapses, an intracellular deposit of insoluble fibrillar proteins (tau protein), and the formation of extracellular senile plaques (amyloid plaques). In most cases and as for cognitive decline (Section 4.1.1), the presence of the ApoE4 gene (in approximately 15% of the population) is a genetic predisposition factor that increases by about four times the risk of developing Alzheimer's disease. For a long time this genetic equipment was indeed said to be the cause of amyloid deposits, and more recently the reduction of cerebral metabolism and hippocampus volume (Liu et al. 2015).

Among the nutrients that may influence the evolution of dementia syndromes, several lipid compounds have already been investigated, and the results deserve to be known by everyone. The main targeted lipids are ω-3 fatty acids (Section 4.2.1), vitamin A and carotenoids (Section 4.2.2), vitamin D (Section 4.2.3), vitamin E (Section 4.2.4), and cholesterol (Section 4.2.5).

4.2.1 ω-3 Fatty acids

In the absence of efficient pharmacological treatments, it is possible to consider for Alzheimer's disease, as for the normal cognitive decline, a therapeutic solution using specific lipid supplies that influence the progression of the disease up to an old age. Through animal models and cultured cells, it has become increasingly clear that a supplementation with EPA and DHA can lower amyloid plaque formation, likely through their protective roles in reducing β-amyloid accumulation via mediating the glymphatic system function in the brain (Ren et al. 2016). Neuroinflammation plays an important role in the onset and also in the advancement of the disease, with ω-3 fatty acids being involved in both the reduction and resolution of inflammation. Several new approaches seem to prove that the origin of these effects may be lipid mediators (oxylipins) derived from ω-3 fatty acids (EPA and DHA) in fish oil (Devassy et al. 2016). Numerous studies have also shown that in addition to its anti-inflammatory role, DHA could improve neuronal survival. α-Linolenic acid could also positively impact Alzheimer's disease.

Considerable research has considered that the late expression of Alzheimer's disease results from an interaction between genetics and environment. These positive developments have naturally lead clinicians to explore the relationships between dietary intake of ω-3 fatty acids and the incidence of the disease. Faced with this developing disease, it seems urgent to define simple and inexpensive therapies applicable quickly and safely as early as possible and preferably in subjects genetically predisposed to the disease.

4.2.1.1 Epidemiological investigations

After the observation of a lower concentration of DHA in the brain tissue in patients with Alzheimer's disease, several epidemiological studies were carried out to make a link between the fatty acid status and the symptom onset. Before 2005, the number and the quality of the published work was not an incentive to relevant conclusions; thus, in 2005, the US Agency for Healthcare Research and Quality made no clear recommendation on the effects of ω-3 fatty acids on cognitive functions and their pathological degradation (http://www.ncbi.nlm.nih.gov/books/NBK11853/#__NBK11853_dtls__).

Since this hardly exciting review, nine large epidemiological studies were performed in a dozen countries. They all established with more certainty that the intake of ω-3 fatty acids of marine origin, attested and estimated by blood concentration, decreased significantly the risks of cognitive decline, dementia, or Alzheimer's disease (Cunnane et al. 2009). Nineteen reliable studies were published between 1997 and 2008 on the same subject. It must be noted that if two of them have shown no effect, the other 17 reported that the risk of Alzheimer's disease increased when the nutritional intake of ω-3 fatty acids decreased. Several other epidemiological studies recently published led to the same conclusions. Very recently, it has been shown that a low level of EPA, but not DHA, in the red blood cells was the best indicator of an increased risk of cognitive performance degradation (Street et al. 2015).

From an analysis of numerous publications, it has been estimated that a large weekly consumption of fish (500 g or more) was associated with a 36% lower risk of suffering from Alzheimer's disease. A statistical calculation allowed for the estimation that for each additional intake of 100 g of fish per week, the risk of developing the disease was reduced by 11% (Wu et al. 2015).

In France, two studies have contributed to the understanding of the effects of this specific diet. The vast PAQUID study coordinated by INSERM in southwestern France has shown a close association between an important and regular consumption of fish and a lower risk of developing dementia or cognitive decline when measured after 7 years (Barberger-Gateau et al. 2002). The second major French study, a part of the COGINUT program coordinated by the National Institute of Agronomic Research, aimed at analyzing the relationship between diet and intellectual performances among the elderly, emphasizing the role of ω-3 fatty acids and antioxidants. This project was developed in 1999 within the Study of Three Cities, including more than 9000 people aged over 65 and studied for 7 years in Bordeaux, Dijon, and Montpellier (Barberger-Gateau et al. 2007). At the end of the program, it seemed that old subjects eating fish at least once weekly, in combination with daily consumption of fruits and vegetables, had a 30% lower risk of developing dementia in the next 4 years, but only in subjects not carrying the ApoE4 gene. The regular use of linoleic acid–rich oils (sunflower, peanut, grape seed) in the absence of a compensation by α-linolenic acid–rich oils (rapeseed or walnut) was associated with a two times greater risk of dementia, but only in noncarriers of the ApoE4 gene. However, it seemed that the regular consumption of various sources ω-3 fatty acids should be associated with an antioxidant supply (fruits and vegetables) for a significantly decreased risk of dementia, with the presence of only one of these healthy eating habits not being sufficient for brain protection. An encouraging support for that approach was recently provided by the demonstration of a reduced incidence of Alzheimer's disease associated with a modified Mediterranean diet (Mediterranean-DASH Intervention for Neurodegenerative Delay [MIND] diet) (Morris et al. 2015b). That prospective study of 923 aged participants living in retirement communities and followed, on average, 4.5 years has suggested that reducing the risk of developing Alzheimer's disease may be obtained with even modest dietary adjustments. For example, the MIND diet score specifies just two vegetable servings per day and both two berry servings and one fish meal per week.

4.2.1.2 *Intervention studies*

It is regrettable that the use of EPA and DHA as dietary supplements has so far provided little in the way of conclusive results, probably because of the variety of the lipids used and a lack of discrimination in the selection of participants in these clinical trials. The interpretation of the positive results obtained with various ω-3 fatty acids raises several questions about their analysis or discussion (Cunnane et al. 2009). Nevertheless, all studies have suggested that the achieved effects were highly dependent on the state of disease development, with these fatty acids being more effective in the early stages of disease. The assessment of the first symptoms (prodromal phase) thus seems essential to ensure the success of this nutritional therapy. This is also true for the current pharmacological or behavioral treatments.

This is the point of view adopted by one of the most renowned specialists of this question, Prof. Greg Cole, of the University of California-Los Angeles, who acknowledged that DHA is able to slow several molecular mechanisms deleterious for the nervous system at the beginning of a disease (Cole and Frautschy 2010). This clinician argues openly in favor of an early nutritional intervention including the administration of DHA and antioxidants, especially in subjects at risk.

These recommendations can only encourage investigators to implement all biochemical, anatomical, and neuropsychological aspects to predict as accurately as possible the risk of transforming a simple state of weak memory into an irreversible dementia. In the absence of reliable markers, it is advisable to follow the nutritional recommendations for lipid intakes by selecting foods that meet the individual requirements.

Several studies have concluded that there was no evidence of any usefulness of a ω-3 fatty acid supplementation to improve the cognitive decline or dementia in the elderly (Dangour et al. 2012). Despite these conclusions, an extensive study carried out in the United States in year 2000 has definitely turned upside down the organization of the future clinical research in this area (Quinn et al. 2010).

This study, among the most serious in recent years, was carried out under the sponsorship of the National Institutes of Health by a group of cooperative research on Alzheimer's disease. This large trial involved 51 research centers gathering together 402 patients divided into two groups: one group received 2 g of DHA daily for 18 months and the group (control) received a mixture of vegetable oils. In addition to numerous psychometric tests, the distribution of ApoE4 gene carriers was investigated.

The overall result of this immense work that lasted several years was a little disappointing because the DHA supplementation led to no detectable benefit with selected tests, but the treatment did not cause unpleasant side effects. Conversely, the noticeable result of this work comes from the comparison of subjects carrying the ApoE4 gene with those who did not. Indeed, only the subgroup without the ApoE4 allele could benefited after an 18-month treatment of a highly significant slowing down of their cognitive decline. It seems, as it has often been reported in other studies, that ApoE4 is a potent regulator agent of the DHA effects on the emergence of dementia (Huang et al. 2005). However, some authors contested these findings, showing that unanimity is not yet fully established for the involvement of that genotype in specific neuropathological risks. These discrepancies may result from several factors affecting the DHA metabolism, such as sex, age, body weight index, and alcohol consumption.

These issues obviously pose a lot of questions because DHA may possibly inhibit the generation of amyloid plaques and slow the cognitive decline in patients lacking the ApoE4 gene, although the intimate mechanisms of this regulation remain unknown (Salem et al. 2015). These results should be related to the recent data indicating that the presence of the ApoE4 gene is the most prevalent risk factor for Alzheimer's disease, because this lipoprotein is expressed in more than half of the patients (Michaelson 2014). The role of the gene coding the protein in the onset of Alzheimer's disease is predominant and indisputable because it is widely associated in the brain, not only with hippocampus atrophy but also with a deposit of amyloid substances (Liu et al. 2015).

It is hoped that future research will target rapidly these regulatory mechanisms without missing the methods of dietary treatment in patients carrying the ApoE4 gene. As it was stressed previously for the role of age (Section 4.1.1), it seems important that everybody could know his or her genetic composition to decide the merits of an early prevention based on DHA supplementation, EPA supplementation, or both without waiting for the first manifestations of a cognitive decline that may lead to an irreversible nervous degeneration. The generalization of this "nutrigenetic" approach seems the most pragmatic, given its relative safety compared to all pharmacological treatments, with their benefits being still very questionable. It must be emphasized that according to a large survey carried out in 2014 by a mutual insurance company (MGEN), 62% of French people have said they are interested in genetic testing and 84% would change their lifestyle upon detection of a genetic defect. It is becoming increasingly clear that a high risk of suffering from a serious cognitive deficiency argues for detecting any specific predisposition to take place as early as possible. That approach is not a bet on probable science progress because solutions already exist to delay or attenuate the manifestations of cognitive decline and dementia. These preventive means involve only a small dietary change or possibly a supplementation with natural products extracted and purified without any chemical modification. It seems evident that genetic tests specific for the cholesterol and fatty acid metabolism would be more widely practiced while remaining faithful to the bioethic laws, because results of these tests could be associated with potential nervous disorders.

Within the framework of the last French Alzheimer plan, a large survey (Multidomain Alzheimer Preventive Trial [MAPT]) initiated by the Toulouse University Hospital (Gérontopôle) was launched in 2008 with 1680 subjects more than 70 years old to determine whether preventive measures, including the administration of ω-3 fatty acids, could protect against memory loss. Everyone hopes to know shortly the outcomes of this vast trial.

Further research will certainly contribute to the discovery of additional therapeutic and especially preventive strategies for the various forms of dementia, and to the knowledge of the biological mechanisms and the treatment modalities convenient at each disease stage. Thus, the future explorations involving ω-3 fatty acids should also consider the recent discovery of Dr. F. Jernerén on the influence of B vitamin (Jernerén et al. 2015). Indeed, in the context of the great OPTIMA (Oxford Project to Investigate Memory and Aging) study, a remarkable new concept was proposed suggesting future developments for a more efficient prevention of Alzheimer's disease. Investigators have shown that in elderly (70 years or older) with mild cognitive decline, ω-3 fatty acid supplementation had a beneficial effect only when there was a joint contribution of B vitamins. Recall that the dietary B-complex vitamins, including B1 (thiamine), B2 (riboflavin), B3 (niacin), B5 (pantothenate), B6 (biotin, folate), and B12 (cobalamin), are important regulators of neurotransmitter function. Vitamin B6, in particular, is an important cofactor for the enzymes that synthesize serotonin, epinephrine, norepinephrine, and γ-aminobutyric acid.

It had also been previously demonstrated that an important intake of B vitamins could slow brain atrophy in this type of patients (Smith et al. 2010). This discovery led the project leader, Prof. D. Smith of the Department of Pharmacology, Oxford University, to state in April 2015 in the *Daily Express* that "it was the first treatment showing that brain alterations associated with Alzheimer's disease may be prevented. This means that simply maintaining a high level of ω-3 fatty acids, associated with a B vitamin supplementation, can greatly reduce the disease risks." Furthermore, Smith believes that this treatment should be administered immediately after the first signs of dementia.

These new data will help in future experiments to form more consistent groups of subjects to avoid result dispersion, thereby increasing their meaning. An alternative could be to supplement subjects with a mixture of B vitamins at appropriate doses before any supplementation with ω-3 fatty acids.

In this area, several new trials already look promising, for example, the LipiDiDiet European project. Indeed, a new dietary supplement (Nutritia Souvenaid®) providing 1200 mg of DHA and 300 mg of EPA daily, vitamins (E, B6, B12), trace elements (selenium), and various metabolites (uridine, choline) was administered to subjects from several European countries who were exhibiting signs of Alzheimer's disease. Well-controlled experiments have demonstrated that this treatment had beneficial effects on memory and patient behavior (Scheltens et al. 2014) and even on the organization of the neuronal networks (de Waal et al. 2014). These trials provide compelling evidence that it is now possible to improve cognitive functions by using a combination of nutrients combining ω-3 fatty acids and several vitamins. Nonetheless, another trial with the same product found no effect on the development of the cognitive decline in subjects clearly affected by the Alzheimer's disease (Shah et al. 2013). As noted, it is likely that the stage of progression of the disease in these two trials is the source of these differences.

Further clinical trials are underway to evaluate all the possible effects of this new product as able to maintain the integrity of synapses, a characteristic that is essential for brain function, on account of ω-3 fatty acids and other compounds.

In the absence of investigations on primary prevention, several basic and clinical studies are needed before considering a widespread prescription of DHA or EPA to prevent or first to treat Alzheimer's disease. These steps forward will benefit from related progress in the molecular diagnosis of the disease, thereby allowing early identification of individuals at risk. They may encourage the support of long-term research with strict protocols to get rapid solutions for fighting against all types of dementia. Pending the outcomes of this future work and considering that the majority of fatty acid supplementations give equivocal results, the main recommendation is again to increase fish consumption, with supplementation of purified ω-3 fatty acids as an alternative. It is regrettable that, in general, academic authorities or care associations that are able to guide government decisions are still very careful regarding support of "nutrigenic" prevention of dementia. In focusing their interest on hypertension or diabetes, these organizations pay, in general, no attention to the influence of essential fatty acids on brain function. However, in later publications, INSERM acknowledged cautiously that "a broad preventive message (more physical activity, better balance diet, a healthy lifestyle) could help to delay the onset or expression of Alzheimer's disease." It is also regrettable that even in the most recent books (Léger and Mas 2015) there is only a discreet mention of a 2003 publication concluding ω-3 fatty acids have beneficial effects on cognitive decline (Heude et al. 2003).

Compared with dementia of neuronal origin, vascular dementia remains a poorly defined syndrome, despite research that began nearly 80 years ago. This pathology is clinically described most often as a cognitive decline related to intracerebral vascular lesions caused by a stroke of ischemic or hemorrhagic origin. Vascular dementia is the second type of common dementia after Alzheimer's disease, being found in 15%–30% of cases; however, this pathology is underestimated in clinical diagnostics, although the risk factors are the same for these two types of dementia. Unfortunately, limited research has been devoted to the effects of ω-3 fatty acids on vascular damage possibly causing dementia. Most epidemiological studies still show an inverse relationship between fish consumption and the incidence of the disease. A large epidemiological study concluded that an increase in fish consumption (from one serving monthly up to four servings per week) significantly lowered by 35% the risk of vascular dementia, but only when consumed fish was not fried (Misner et al. 2001). Conversely, the important Italian study Gruppo Italiano per lo Studio della Sopravvivenza nell'Infarto Miocardico (GISSI)-Prevenzione was unable to detect any effect after a daily supplementation of 850 mg of ω-3 fatty acids. The few in-depth studies and the small number of dementia cases reported in each of these studies do not yet allow appropriate therapeutic options to be determined. Further work on a larger scale is needed to explore the complexities of vascular dementia, which is usually difficult to diagnose with certainty.

Globally, 15 projects are underway or are about to be completed in various research centers that have focused on the neuropsychiatric effects of ω-3 fatty acid supplementation in patients suffering from more or less pronounced dementia (http://clinicaltrials.gov). The prospective research should now consider the ApoE polymorphism because it seems to closely influence the DHA benefits either in epidemiological or experimental investigations.

4.2.2 *Vitamin A and carotenoids*

4.2.2.1 *Vitamin A*

The importance of vitamin A for the nervous system was revealed in 1935 after the discovery of its role in eye function. Later, it was shown that the brain needed an active form of vitamin A called retinoic acid not only for the control of cell proliferation during brain development but also for neuroplasticity, a complex mechanism allowing neurons to alter their connections with other cells. Evidence from studies of animal models suggests that vitamin A is essential for the cognitive functions of the brain (Stoney and McCaffery 2016).

Around 2000, several studies performed with animals highlighted the fundamental role of vitamin A (Section 7.3) in the development and the function of the adult brain. All of these studies emphasized the deleterious effects of a vitamin A deficiency on the hippocampus, one important area for learning and memory (Misner et al. 2001). These outcomes have led researchers to note this societal problem worldwide, because it involves the mental health of hundreds of millions of adults and children deficient in vitamin A.

Several studies have shown that dysregulation in vitamin A supply was well related to the onset of Alzheimer's disease (Goodman and Pardee 2003). Such a conclusion was confirmed by animal experiments establishing a close link between vitamin A deficiency, loss of various memory types, and especially appearance in the brain of β-amyloid peptide, an indisputable marker of Alzheimer's disease (Sodhi and Singh 2014). Conversely, a vitamin A treatment of these animals was able to reduce the accumulation of this peptide and also improve memory performances (Ding et al. 2008). Specific research carried out in vitro was able to clarify the molecular mechanisms leading to these effects (Ono and Yamada 2012).

Other observations were added to the clinical description of the disease, such as low vitamin A blood content in patients with the Alzheimer's disease, with the results being more significant in the late form (or sporadic form) of the disease. It is possible that much of the effects of vitamin A (or retinoids), as carotenoids, could be attributed to their antioxidant actions, thus fighting against the formation of free radicals known to be highly toxic for brain tissue (Smith 2006).

Unfortunately, clinical studies are desperately needed in this field, with only one being officially launched worldwide in 2012; this study used a form of vitamin A prescribed for the treatment of psoriasis (Acitretin). Trials are under way using a synthetic analogue of vitamin A, Targretin® (bexarotene). This retinoic X receptor agonist was shown to decrease brain amyloid formation and to improve synaptic and cognitive functions in animal models of Alzheimer's disease. The treatment of one patient with mild Alzheimer's disease (6 months with 300 mg bexarotene/day) induced a 40% memory improvement and a 20% decrease of the tau protein in the cerebrospinal fluid, whereas no significant side effects were noticed (Pierrot et al. 2015). This observation indicates that bexarotene may improve memory performance and biological markers at an early stage of Alzheimer's disease. It is hoped that ongoing clinical trials will provide much more consolidated data to validate that vitamin A analogue as a potential drug in the treatment of neurodegenerative diseases.

In conclusion, a therapy with vitamin A should lead to major benefits in the evolution of the Alzheimer's disease, particularly by modulating the β-amyloid synthesis, a step that is not the target of the current treatments. Moreover, all results support the beneficial role of retinoids through their antioxidant and antiapoptosis activities and by their role in cellular differentiation and activation of cholinergic functions. Retinoids may be an important repair factor for the nervous system by activating cell protection and regeneration.

In the absence of other information, the issue can once again be raised regarding the lack of interest of the pharmaceutical industry for medications producing little return on investment. In future research, it should be necessary to better understand the involvement of vitamin A in the signaling pathways and possibly the oxidative reactions within the brain. These pathways are now well known to play an active role in the genesis of pathological lesions in patients with the Alzheimer's disease.

4.2.2.2 Carotenoids

As for vitamin A, it was frequently observed that patients with Alzheimer's disease had lower carotenoid blood levels than healthy subjects. Among carotenoids (Section 7.3), lutein and zeaxanthin displayed the most important decreases. This relationship was also observed in the retina (macula), wherein the progression of the disease can be monitored by a photometric examination of the ocular fundus. All these outcomes are consistent to explain the higher frequency (about two times) of AMD in patients with Alzheimer's disease (Nolan et al. 2014).

Thus, carotenoids seem to be very active agents in lowering manifestations of Alzheimer's disease, and future research could lead to the proposal of a natural means to fight against this disease, likely by starting dietary supplementation very early. Pending more efficient and more specific treatments, the regular consumption of vegetables rich in carotenoids may currently be considered an effective prevention.

4.2.3 Vitamin D

Historically, vitamin D is considered responsible for calcium homeostasis and thus key for bone metabolism. Gradually, over the past 30 years, physiologists have shown that this vitamin also has functions in cell proliferation and differentiation, immunity, and many other metabolic systems (Leray 2015).

Long after its discovery in 1920 by the American chemist E. V. McCollum, vitamin D gained notice in neurochemistry, soon after the synthesis in the 1970s of its active form, calcitriol, and especially after the discovery in

1987 of its receptor in the brain. In the past 15 years, the importance of this lipid vitamin in brain function has continuously grown to the point where it is now considered as a hormone and classified among neurosteroids (McGrath et al. 2001).

As with all vitamins, progress came from the study of deficiency states in animals and in humans. Recently, it became increasingly clear that levels of vitamin D that were too low in blood of adults were correlated with serious neuropsychiatric problems and sometimes with symptoms of neurodegenerative diseases (Eyles et al. 2013). In terms of brain structure, the observation of an abnormal increase in the white substance volume, measured by MRI in the vitamin D–deficient elderly, characterized their motor and cognitive decline (Annweiler et al. 2015a). This observation is fully consistent with that of an inverse relationship for the gray matter (Brouwer-Brolsma et al. 2015).

These findings can only suggest to correct any vitamin D deficiency, considering the importance of the induced pathologies and the extent of the population concerned, not to mention the huge financial consequences.

The numerous studies conducted in animals have all led to the conclusion that vitamin D plays the role of a neuroprotective actor, able to fight against the age-related cognitive decline in improving learning and memory (Latimer et al. 2014). Even more interesting and yet known for nearly 30 years is that vitamin D can accelerate acetylcholine synthesis in some brain areas. A low synthesis of this neurotransmitter is recognized as fundamental in the development of dementia (Sonnenberg et al. 1986). The recent demonstration of control of serotonin biosynthesis by vitamin D provides insight into the modulation of behavior and cognitive functions by this vitamin (Patrick and Ames 2015). In addition, research on animal models has demonstrated that vitamin D not only improves synaptic plasticity but also reduces amyloid deposits. This knowledge makes it possible to hypothesize that vitamin D may be effective in patients at risk or with an Alzheimer's disease in progress.

4.2.3.1 Epidemiological investigations

Nearly 100 trials were carried out in the past 10 years to explore the potential role of the vitamin D status on the cognitive decline leading to dementia and more specifically to Alzheimer's disease. These investigations consisted either of observing a group of several subjects (a cohort) during a short time (observational studies) or of following a population for the longest time possible (prospective studies). In both cases, the results generally led to the conclusion that subjects with vitamin D blood levels lower than the current recommendations (less than 30 ng/mL or 78 nmol/L) had a greater risk of developing a dementia condition than subjects with higher levels (equal or greater than 30 ng/mL).

An example of a typical observational study is that of Llewellyn et al. (2009) at the University of Cambridge, UK. They selected 1766 British subjects of both sexes, aged 65 years or more (mean age 80 years). A blood vitamin D determination and a psychometric test similar to the MMSE (Section 7.9) were processed in each participant. Statistical analysis of all the results showed that subjects having a vitamin D concentration in the range of 17–25, 12–17, and 3–12 ng/mL had a risk of cognitive disorder multiplied by 1.1, 1.4, and 2.3, respectively, compared to subjects with a vitamin D level greater than 25 ng/mL.

This work proved that a decreased vitamin D level in blood was associated with an increased risk of developing dementia, with the more deficient subjects having a two times higher risk than control subjects of the same age group.

To illustrate an example of a prospective study, the work done at the University of Minnesota, Minneapolis, is considered (Slinin et al. 2012). This study followed 6257 female participants, with a mean age 77 years, for 4 years. Besides vitamin D determination in the blood, they proceeded with a test of global (MMSE) and executive (TMT) function (Section 7.9). At the onset of the study, the authors found an association between vitamin D and cognitive impairment similar to that detected in the previous study. After 4 years, a new psychometric examination in the same subjects showed that those initially deficient in vitamin D (concentration less than 10 ng/mL or 26 nmol/L) had a risk of cognitive disorders two times higher than that of the subjects having levels above that threshold.

The study of D. L. Llewellyn, comparable to the previous study, provided finally comparable results after 6 years: the risk of a significant decline (3 points on the MMSE test) in subjects with vitamin D blood levels below 10 ng/mL was 30% higher than in subjects having a level above 28.5 ng/mL (Llewellyn et al. 2010). A recent cohort study (Littlejohns et al. 2014) in the United States by the same team involving 1658 elderly people monitored for a mean of 5.6 years found an incident Alzheimer's disease for those with vitamin D <25 nmol/L two times higher than for those with vitamin D >50 nmol/L. Similar results were obtained after following more than 10,000 Danes during 30 years (Afzal et al. 2014); 412 Koreans aged over 65 years for 5 years (Moon et al. 2015); and 382 Americans with a mean age of 75.5 years (Miller et al. 2015).

These examples illustrate clearly the hypothesis prevailing among specialists of an adverse effect of vitamin D deficiency in maintaining the cognitive status during aging, a situation that may even lead to dementia. It is necessary to emphasize that although most of the elderly have a vitamin D deficiency, not all will develop Alzheimer's disease. Hypovitaminosis cannot alone explain the occurrence of the disease, thus suggesting that a vitamin D supplementation may not be sufficient in each case for an effective prevention.

To broaden the field of study and open discussion, an awareness of two recent reviews of published communications on this subject over the past decade is needed.

The first review, conducted at the Department of Geriatrics at Utrecht University, The Netherlands, selected 28 communications combining the observations on nearly 57,000 people (van der Schaft et al. 2013). The review revealed that almost 72% of the work found an association between low vitamin D levels and cognitive decline, thus indicating a higher risk of dementia. But no study showed any negative effect for a high vitamin level.

The second review was done at the Department of Geriatrics, Angers University, France (Annweiler et al. 2013). The results of 14 research programs confirmed the previous findings indicating that the executive functions are particularly sensitive to the plasma vitamin D concentration. Subsequent to that final review, the same team has shown that a vitamin D deficiency (serum concentration <28 ng/mL) was significantly associated with more subjective memory complaints (Tot Babberich Ede et al. 2015). Recall that memory complaint is a light but frequent subjective issue in normal aging because it affects up to half of subjects aged over 50 years. Recent findings have shown that the vitamin D threshold associated with cognition issues is likely around 10 ng/mL. Subjects with concentrations lower than 10 ng/mL have important risks of cognitive disorders, whereas the risks become lower when the concentrations are higher than 30 ng/mL (Annweiler 2016).

A meta-analysis of five more recent epidemiological surveys confirms that vitamin D deficiency increases of more than 20% the risk of developing dementia or Alzheimer's disease (Shen and Ji 2015).

These studies are clearly unable to definitively establish a causal relationship between vitamin D and cognitive decline. Many factors can be invoked to weigh the results, but the overall trend of a correlation between the two settings does not seem to be challenged, despite the diversity of the analytical approaches. When asked whether hypovitaminosis D precipitates cognitive disorders or vice versa, Dr. Annweiler (Annweiler 2016) answered: "Longitudinal prospective studies have provided possible answers by showing that older ethnically diverse groups of individuals with hypovitaminosis D have a significantly increased risk of cognitive decline, Alzheimer's disease, and all-cause dementia, compared to those with higher vitamin D concentrations." No relationship has been reported after some trials, but a close examination of the conditions of these tests easily reveals that several uncontrolled biases often distort the conclusions. Indeed, authors emphasize frequently the difficulty of achieving global analyzes from several investigations, considering the diversity of the measurements and the psychometric tests. It must be remembered that the transition between a subjective cognitive complaint and a characterized Alzheimer's disease is progressive and that no specific biological marker can be used to characterize any intermediate step. Moreover, the most recent large French study (SU.VI.MAX 2, Supplementation in Antioxidant Vitamins and Minerals) has shown that positive effects of vitamin D on some cognitive performances were demonstrable only in uneducated subjects (Assmann et al. 2015).

The importance of the contribution of genetic factors in the modulation of cognitive functions, as for the vitamin D status, must not be underestimated (Maddock et al. 2015). Whereas individuals carrying the ApoE4 gene may have some risk having reduced cognitive performances (Section 4.2.1), they show higher circulating vitamin D levels (Egert et al. 2012). Unexpectedly, these findings are more numerous in black-skinned people and are found even more frequently in the highest latitudes. Much work has also emphasized the importance of the genetic polymorphism of the vitamin D receptor in connection with a premature neuronal aging.

These complex relationships between genes and environment, yet poorly explored by clinicians, are undoubtedly relevant to the evolution of our species, with their knowledge also being essential to develop and improve new therapeutics to fight efficiently against Alzheimer's disease.

4.2.3.2 *Intervention studies*

For vitamin D, as with all vitamins or hormones, the study of the causal links between that compound and the physiological effects requires precise intervention studies. The experimental design of these trials should include a random distribution of subjects between a control group receiving no treatment or a placebo and an experimental group supplemented with a defined dose of vitamin D precisely administered. At the end of the experiment, the evaluation of both groups is carried out using biochemical analysis combined with a neuropsychological assessment of each subject. That approach, however, presents some limits for the interpretation of results because the protocol is often designed in well-controlled experimental conditions, but it does not reflect a diverse population commonly encountered on the ground.

Little experimental work has been done to date to elucidate the effects of a correction of too low vitamin D levels on dementia. This lack is probably a consequence of the difficulty of implementation of these trials and their cost. The very low added value of the product may also contribute to the scarcity of such research in the pharmaceutical field.

The first attempt was made in 2008, but the authors have not reported any positive effect on cognitive performances and behavior in 25 nursing home residents aged about 86 years and deficient in vitamin D (Przybelski et al. 2008). No doubt, this is on the account of the use of ergosterol (vitamin D_2), a plant analogue of cholecalciferol (vitamin D_3) known to be biologically less active. A 4-week trial seems also too short to observe significant changes in the brain.

A large trial was undertaken with nearly 4000 American women of average age 70 years, half of which were supplemented with 400 IU/day of vitamin D (Rossom et al. 2012). After a 3-year treatment, no advantage over cognitive decline and dementia incidence could be found. The use of an insufficient amount of vitamin D, a small number of reported neuropsychological incidents, and the unique selection of women probably explain this lack of effect.

The most encouraging trial was reported by Annweiler et al. (2012a). Twenty patients, with a mean age of 82 years, received a supplementation of 800 IU/day of vitamin D for 16 months. During the study, the blood level of vitamin D increased from 6 to 11 ng/mL, and all the psychometric tests displayed improved global and executive cognitive functions. Despite the use of small cohorts, this work confirmed the assumptions made from the epidemiological studies summarized above: it helped to demonstrate that dementia does not cause a hypovitaminosis, but rather the reverse.

Future treatment of cognitive decline and dementia therefore needs the use of a section including vitamin D. The vitamin supply may be carried out through the diet, as a nutritional supplement or possibly by its natural synthesis during prolonged sun exposure, although that last source is recognized as inefficient in the elderly. Such therapy may be also considered in conjunction with conventional treatments, because Annweiler et al. (2012b) has shown that the efficiency of a treatment with memantine, a dementia symptomatic treatment, was potentiated by the addition of vitamin D.

Other recent data reinforce the hypothesis of a role of vitamin D in the maintenance of cognitive functions. Among them, are noteworthy the high amounts of the vitamin D transport protein in the cerebrospinal fluid in individuals with Alzheimer's disease. The transport protein status may thus contribute to decrease the cellular availability of vitamin D.

For ω-3 fatty acids, research is still necessary before proposing a vitamin D treatment in healthy individuals to prevent illness or to limit the worsening of preexisting cognitive issues. It is desirable that the previously published work can used actively to encourage more researchers to explore the various aspects of vitamin D activities.

All critical aspects between vitamin D and cognitive abilities were analyzed in early 2013 by a group of 12 international experts grouped as a task force focusing on key questions related to the role of vitamin D in Alzheimer's disease and related disorders (Annweiler et al. 2015b). The following questions were addressed:

To the questions:

- Can hypovitaminosis D be considered as a risk factor for cognitive disorders or Alzheimer's disease (decline and dementia)? All experts answered yes. The experimental evidence for an action of vitamin D on the brain is clear enough at the present time to justify that opinion.
- Can serum vitamin D status be considered a useful biomarker for the diagnosis of Alzheimer's disease and related disorders? All experts except one agreed to answer no. The main argument is that vitamin D deficiency is too common in the elderly and is not specific for any cognitive disorder.
- Can serum vitamin D status explain part of the variability in symptoms of Alzheimer's disease and related disorders in older adults? All experts agreed that vitamin D can explain, at least in part, the diversity of symptoms. Hypovitaminosis D affects many organs other than the brain and has been associated with numerous diseases. These conditions may be found in patients with both hypovitaminosis D and Alzheimer's disease and related disorders, thus affecting a patient's functional independence and the clinical status.
- Should vitamin D supplementation be part of the care management of older adults with cognitive disorders or with Alzheimer's disease and related disorders. All experts except one agreed to answer yes. The main argument was that the clinical results are generally favorable. Indeed, vitamin D has positive effects on the brain, but also on many other organs including bones. In addition, this supplementation is without risk.

Why did the health ministers in several countries not impress more upon their agri-food industries the need to speed the enrichment of most common foods (dairy products, bread, cereals, oils), enrichment that was already allowed in many countries, as in France, since the 2006 law (Leray 2015)? Even if this allows only a maximum intake of 5 μg/day of vitamin D. It seems now appropriate to take into consideration the official scientific studies that revealed that the average vitamin D intake was corresponding to half of that value as in France, hence confirming the predominance of a deficiency state in several populations.

Furthermore, one can manifest astonishment when the French High Authority for Health published in 2013 that the usefulness of blood vitamin D analysis is not demonstrated. That authority has retained only partially the osteoporosis pathology. In 2014, the National Health Insurance decided to abolish the reimbursement of expenses associated with these analyses.

The same high authority has curiously also acknowledged that a vitamin D supplementation could be established and maintained without a prior biochemical determination.

Despite the weight of the scientific information, it is surprising that health authorities do not presently envisage some preventive measures encouraging vitamin D supplementation, at least for populations at risk. Indeed, that kind of decision could delay by at least by 1 year the appearance of any dementia if individuals over 65 years were supplemented. The total prevention cost (approximately €140 million for 100,000 IU every 2 months) would be amply compensated by cost savings estimated to about €2 billion for Alzheimer's disease (on an annual total of €16 billion), but also for those related to fractures, infections, and many chronic diseases (Leray 2015).

Globally, 19 projects that are declared completed were still recruiting in 2016 in various research centers targeting the neuropsychiatric effects of vitamin D in patients suffering from various types of dementia, including Alzheimer's disease (http://clinicaltrials.gov).

In conclusion, with an increasing number of people affected by cognitive decline usually leading to Alzheimer's disease, and with the broad spectrum of activity of vitamin D, it seems appropriate to undertake a large deficiency screening plan for this vitamin. The expected results will be clearer if the screening is carried out as soon as possible. This plan must be followed by specific advice to restore vitamin levels greater than 30 ng/mL blood through diet, sun exposure, or possibly with the intake of pharmaceutical supplements. Despite uncertainty about the most effective doses and the influence of the genetic terrain of individuals, it is imperative to raise the awareness of people on the importance of vitamin D, especially if one estimate that 80% of adults are now insufficient (10 ng/mL < blood levels <30 ng/mL) or deficient (blood levels <10 ng/mL) in the majority of countries is accurate. The focus should therefore be on the need for a vitamin D daily intake consistent with the recommendations of medical authorities (1000–1500 IU/day after 50 years old, Section 7.4). This is particularly important for populations recognized as deficient in vitamin D, such as the elderly, young children, and people with much pigmented skin or living mostly out of direct sunlight.

4.2.4 Vitamin E

The term "vitamin E" is not equivalent to α-tocopherol, despite the large amount of work devoted to that chemical form (Section 7.5). The other forms belonging to the vitamin E group must not be ignored because they are also the subject of an increasing number of investigations in the brain area. It is significant, for example, that barely 1% of the literature on vitamin E concerns the tocotrienols.

The importance of vitamin E for the central nervous system seemed evident as early as at the beginning of the twentieth century when H. M. Evans and G. O. Burr described the presence of paralysis in young rats born from vitamin E–deficient females.

Vitamin E is the most abundant lipid antioxidant in the human body, with that property probably explaining its well-known protective effects on biological membranes and therefore nerve cell membranes. These effects are also potentiated in the presence of vitamin C that regenerates α-tocopherol after its inactivation by free radicals, a property explaining the joint use in some experiments of both vitamins.

The frequent demonstration of a close association between oxidative stress, aging, and some nervous pathologies, such as Alzheimer's disease, has emphasized the possible use of antioxidant compounds in the prevention and treatment of these areas. Therefore, antioxidants may be also useful in the fight against memory loss and dementia, because they have proven to be beneficial for the cardiovascular system (Leray 2015), itself involved in cognitive disorders.

4.2.4.1 Epidemiological investigations

It was not until 1999 that a link between vitamin E (and not vitamin C or β-carotene) and memory performances was clearly established in considering a vast multiethnic US population (Perkins et al. 1999). A year later, work including blood analysis in many centenarians confirmed this information (Klapcinska et al. 2000). The hypothesis of a possible protective effect of vitamin E on cognitive capacities could be issued.

Here, only the best-documented effects of only some components of the vitamin E complex on cognitive decline and dementia are reported, with further data summarized in a review by La Fata et al. (2014).

Many studies based on dietary surveys have attempted to reveal a possible association between antioxidants and memory. The specific contribution of the dietary vitamin E remains difficult to estimate, because it is regularly ingested with many other natural antioxidants, such as vitamins A and C, carotenoids, and several polyphenols. The role of each compound and the importance of the mixtures remain undetermined. Among the epidemiological studies open to debate, a US survey conducted in 2002 followed nearly 2900 individuals aged 65–102 over 3 years (Morris et al. 2002). The statement of the vitamin E intake and the use of four different psychometric tests allowed to correlate the high doses of vitamin E with a reduced cognitive decline. A similar study conducted in The Netherlands in 1996 (Rotterdam study) did not reveal any correlation. The extension of that study for 6 years on the risk of dementia and Alzheimer's disease has nonetheless revealed a modest, but significant beneficial effect of vitamin E, with that effect being more pronounced in smokers. Another extension of the same study for 10 years has confirmed these results, eliminating the interferences of the β-carotene and vitamin C effects (Devore et al. 2010). The large French SUVIMAX2 study came to similar conclusions after a follow-up of 10 years (Peneau et al. 2011).

These examples show the difficulty of these approaches, facing a time scale often not compatible with investigations requiring a wide financial support. Furthermore, a possible reason to justify the inconsistencies is the multifaceted feature of Alzheimer's disease with a progressive development and different levels of severity.

A Swedish team conducted a clinical examination of patients with moderate-to-severe cognitive disorders (Alzheimer's disease) by performing MRI investigations in the brain and determination of various circulating forms of vitamin E (Mangialasche et al. 2013a). The authors concluded that it should be possible to use serum vitamin E as an indicator of the severity of disorders, as the vitamin concentration was lower when the disease was more pronounced. They also noted, throughout a period of 8 years that older subjects with elevated blood levels of γ-tocopherol and tocotrienols developed less cognitive impairments (Mangialasche et al. 2013b). The importance of γ-tocopherol seems to be supported by the discovery of a link between its high brain level and a reduced deposition of amyloid plaques between neurons and neurofibrillary tangles within the neuronal cells (Morris et al. 2015a). These results demonstrate the complexity of the studies in the elderly dealing directly with the prevention of cognitive decline, but they should already be encouraging people to diversify their dietary plant sources to ensure a regular intake of the different vitamin E forms (Section 7.5).

Precise measurements of vitamin E could thus help to establish a diagnosis that remains still difficult to provide. Although the link between the vitamin E status and the cognitive decline level is still unclear, it is interesting to note that the blood vitamin E level does not seem to be connected to a patient's nutritional status (Lopes da Silva et al. 2014). This could remove from the debate a dietary default in subjects known to be often struck by malnutrition.

4.2.4.2 Intervention studies

Early research on the joint evolution of the vitamin E status and the severity of cognitive impairment have led some clinicians to evaluate the effect of a vitamin supplementation on these disorders. Studies using moderate doses of vitamin E have failed to demonstrate any effect on cognitive performance; however, doses up to 1340 mg/day have yielded positive results in one laboratory, but not in another (Farina et al. 2012). It took still 20 years to detect a beneficial effect of a regular supplementation with vitamins E and C on cognitive function after examining nearly 15,000 aging women (Grodstein et al. 2003). In a randomized trial conducted in the United States, a high daily dose of vitamin E resulted in a slowing of the functional decline in patients with a mild-to-moderate Alzheimer's disease and was also effective in reducing caregiver time in assisting patients (Dysken et al. 2014). In the group supplemented with α-tocopherol, a delay in the annual rate of clinical progression of 19% or approximately 6.2 months was observed over the follow-up period (5 years). The discrepancies observed between these investigations cannot be explained; however, it is possible that the controversy over a beneficial effect of vitamin E in Alzheimer's pathology is based on the observation of positive effects only when they are accompanied by a decrease in the oxidative stress. In contrast, the refractory subjects who experience no change with that antioxidant have an unfavorable evolution of their cognitive status, even faster than in subjects receiving a placebo (Lloret et al. 2009).

The comparison between studies done in different countries is often made difficult by the various chemical forms of the vitamin ingested naturally by patients. Thus, Europeans mostly absorb α-tocopherol from sunflower and olive oils, whereas Americans absorb mainly γ-tocopherol from soy and corn oils. Similarly, for intervention studies, the administered form is mainly α-tocopherol found commonly on the market worldwide, but until now no studies have considered the other forms, tocopherols or tocotrienols, isolated or as mixtures. Research undertaken on cellular models have already provided sufficiently encouraging results to consider, in the near future, the development of effective treatments involving specific forms of vitamin E that have not as yet been explored in clinical research. In addition, results obtained in animals suggest that sesame oil consumption could strengthen the effects of vitamin E on slowing nervous degeneration processes, a combination likely at the origin of the low development of Alzheimer's disease in the Indian population.

In conclusion, the positive effects obtained in several studies and the relative safety and low cost of its supplementation suggest vitamin E as a nutritional compound to promote healthy brain aging and to delay Alzheimer-related functional decline. Despite the still discordant results provided by clinicians (La Fata et al. 2014), it seems wise to maintain a permanent dietary vitamin E intake at least equal to the recommendations made by the medical authorities (12 mg/day for an adult) (Section 7.5). That daily intake is easily covered by consumption in moderate amounts of vegetables, such as spinach, cabbage, pepper, and pumpkin, but also meat and fish. An addition of oil, if possible from corn or seeds (almond, hazelnut, peanut) may achieve much higher vitamin E levels. For therapeutic doses of vitamin E, such as those used in some work, plant sources are no longer suitable, and it is necessary to use commercial vitamin supplements. These preparations are either of chemical origin, containing only one form of vitamin E, or of natural origin, mostly as concentrates from palm oil refining. Vitamin supplements have the advantage of containing high concentrations of almost all vitamin E forms (Leray 2015).

4.2.5 Cholesterol

The brain is the richest organ containing cholesterol (Section 7.6). Although it is about 1/50th of the body weight, it contains about one quarter of all the body's cholesterol amount. Brain cholesterol, which exists in a free state, mainly occurs in myelin, astrocytes, and nerve cell membranes. It has long been known that nerve cells synthesize their own cholesterol, being virtually independent of blood supply (Morell and Jurevics 1996). Brain cholesterol is thus mainly synthesized in situ, but mainly in astrocytes and oligodendrocytes. In contrast to developing neurons, mature neurons can synthesize only a small amount of cholesterol. It is well-known that cholesterol plays a major role in the properties of neuronal membranes by modulating the functions of enzymes, receptors, and ion channels.

Since the discovery of the cholesterol transport protein ApoE and the potential links between cholesterol and Alzheimer's disease, major efforts have been devoted to that problem (Canevari and Clark 2007).

Animal studies have shown that a cholesterol-enriched diet is almost always associated with a deposit in the brain of an Alzheimer's disease marker, the β-amyloid peptide; conversely, the inhibition of cholesterol biosynthesis reduces the production of that peptide. For the first time in 1994, a possible link between cholesterol and Alzheimer's disease was proposed in rabbits (Sparks et al. 1994). Later, other work carried out using animal models for the disease found that a cholesterol-rich diet caused not only atherosclerosis, as expected, but also a learning disability in space perception (Li et al. 2003). One of these models lead to stronger evidence of a causal link between the neuronal accumulation of cholesterol and that of the tau protein, a marker of Alzheimer's disease (Burlot et al. 2015).

In addition, it was often stressed that the appearance of atherosclerosis in humans was frequently accompanied with neuropsychological disorders, thus resulting in a poor oxygenation of the nervous areas mainly involved in memory mechanisms. That harmful role of cholesterol on memory received a beginning of an explanation by noting it was correlated with a loss of dendrites and cholinergic dysfunction, and with increased inflammation (Granholm et al. 2008). All these disturbances clearly evoke some typical effects detected in Alzheimer's disease.

Although the relationships between blood cholesterol and disease onset are poorly defined, there is no doubt that cholesterol promotes the genesis of the β-amyloid protein whose metabolism is controlled by the amyloid precursor protein (Allinquant et al. 2014). This complex field is being explored and new strategies for prevention or treatment of the neuronal degeneration will emerge within a short time.

The various assumptions involving cholesterol and dementia are mainly the result of epidemiological studies attempting to link the incidence of the disease to high cholesterol levels (Xue-shan et al. 2016). The extremely widespread use treatment of hypercholesterolemia with statins has prompted clinicians to explore the possible effect of the inhibition of cholesterol synthesis on Alzheimer's disease incidence.

In epidemiological investigations, many studies have shown that high cholesterol levels are harmful for learning and memory and also contribute to the development of MCI and Alzheimer's disease. That was the conclusion arising from the large Finnish study led by Dr. M. Kivipelto, a specialist in this area at the University of Kuopio (Solomon et al. 2007). The authors of this important work, spread over 21 years, have followed nearly 1500 old persons living in the eastern part of Finland. First, they found that, compared to control subjects, cholesterolemia was higher in people with light cognitive impairment and even higher in subjects with dementia. Indeed, it became clear that a high cholesterolemia (>2.5 g/L) around the age of 50 years represented an increased risk of greater cognitive impairment in later life. Conversely, a lower cholesterolemia between 50 and 70 years was often accompanied by an aggravation of neuropsychological disorders. These observations are problematic to understand. Are the disturbances a reflection of a lifestyle change during advancing age? This is unlikely because low cholesterol levels were also recorded in members of aging twin pairs having a severe cognitive decline at approximately 56 years of age (Swan et al. 1992). It is also possible to hypothesize that low cholesterol levels may be related with other pathologies associated with poor cognitive performances. Pending further clarification, it seems wise to ensure and maintain around the age of 50 years a total cholesterol level lower than 2 g/L in agreement with the medical authorities (Leray 2015), to reduce the risk of cognitive decline at an advanced age.

All research in this area has not led to the same conclusions; some research has found no relationship between blood cholesterol and memory disorders. After detailed analysis of negative reports, it seems that these studies focused on the very elderly, whereas studies reporting a positive correlation were performed in younger subjects. This relationship is consistent with the observation of a direct correlation between cholesterol and the presence of amyloid deposits identified in the brain at autopsy, but in subjects under 55 years old only. Through the Maine-Syracuse study, Dr. G. E. Crichton showed that although total cholesterol was not correlated to the cognitive function of people over 60 years, the situation was different for the "good" cholesterol, included in the high-density lipoprotein (HDL) (Crichton et al. 2014). In fact, subjects aged 60–98 years having an acceptable HDL cholesterol level (>600 mg/L) had the highest scores for several psychometric tests; that was not the case for total cholesterol, low-density cholesterol, cholesterol, or triglycerides.

It is obvious that any causal relationship between cholesterol and dementia is far to be demonstrated, because the cerebral metabolism of this lipid is still poorly known. The relationships between the blood compartment and nerve cells, with these being partially protected by the blood–brain barrier, are also not well understood. These reserves are strengthened by the demonstration of a close connection between the inhibition of the neuronal cholesterol synthesis and the resistance to growth factors associated with an increase of apoptosis in the presence of β-amyloid peptide (Fukui et al. 2015). All these effects could contribute to accelerate the neurodegeneration processes, such as those characterizing the Alzheimer's disease.

These results suggest a very relative importance of the circulating cholesterol, related directly to the dietary cholesterol, in the deposition of β-amyloid peptide and thus in cognitive impairment.

The problem is even more complex because many studies have confirmed the hypothesis of an important role of oxidized derivatives of cholesterol in the development of dementia and the progression of the Alzheimer's disease (Gamba et al. 2015). Several studies have emphasized the involvement of these derivatives in brain inflammation, accumulation of β-amyloid peptide, and neuronal death.

The widespread treatment of hypercholesterolemia with statins has pushed epidemiologists to examine the effects of these drugs, largely used worldwide, on age-related cognitive decline and the onset of dementia.

Prof. A. Solomon of the University of Kuopio confirmed previously achieved outcomes in studying nondemented patients, taking or not taking statins during almost 21 years. The conclusion was that a high blood cholesterol level around 50 years old led to a high risk of cognitive decline observed 20 years later, with these risks being nonexistent in patients treated with statins (Solomon et al. 2009). But it must be remembered that beneficial effects of statins on cognitive performances were often reported in older but nondemented subjects (Yaffe et al. 2002). The mechanisms involved are poorly understood, but in addition to the inhibition of cholesterol metabolism, statins may have many other targets in the brain (Crisby 2006). Among the benefits, there is a lowering of the reduction of β-amyloid production, a peptide known to be at the origin of neuronal degeneration (Carlsson et al. 2008). Similarly, the administration of probucol, another cholesterol-lowering drug, to subjects with a mild-to-moderate sporadic Alzheimer's disease led to a significant increase of ApoE levels and to a decrease in the β-amyloid and tau protein concentrations in the cerebrospinal fluid (Poirier et al. 2014). These changes were also correlated with an improvement in cognitive performances.

All this initial work has prompted clinicians to use statins in the treatment of neurodegeneration diseases such as Alzheimer's disease. Unfortunately, the results are still very controversial, but some works suggested a long-term beneficial effect on the development of memory disorders and even on the progression of dementia. Thus, a large meta-analysis has shown that the temporary or continuous use of statins would be responsible for a reduction of 48% or 76%, respectively, of the risk of developing Alzheimer's disease (Xu et al. 2015). However, other studies indicated an absence of effect. A few studies have even set focus on the possibility of harmful effects. For example, in 2014 the US Food and Drug Administration advised doctors and statin users to use great caution in their treatment and that all patients should be cautioned about the possibility of memory loss or confusion at all ages and for all forms of statins. Although these symptoms are not severe and are reversible, the problem of the effects of these inhibitors of cholesterol biosynthesis on brain function and Alzheimer's disease remains open (Wanamaker et al. 2015).

Statins undoubtedly have a pharmacological interest in the incessant fight of medicine against atherosclerosis, but their side effects in the brain are at present poorly understood and therefore should make patients cautious. It is hoped that the research developments in this area may quickly provide valuable insight for preventive treatments of important degenerative diseases. Another way of investigation was opened by researchers who explored the effects of plant sterols, lipids naturally included in our diet. These sterols, known as inhibitors of cholesterol absorption, are able to cross the blood–brain barrier, to accumulate in nerve cell membranes and to reduce the generation of the β-amyloid peptide in cultured cells (Vanmierlo et al. 2015). Among these sterols, the most therapeutically convenient could be brassicasterol and sitosterol, with their concentrations being lower in the cerebrospinal fluid of patients with Alzheimer's disease than in healthy subjects (Vanmierlo et al. 2011).

Globally, 22 clinical projects are about to be completed or were still recruiting in 2016 in various research centers targeting the importance of cholesterol in patients suffering from more or less pronounced dementia (http://clinicaltrials.gov).

In conclusion, it seems prudent to strictly observe the recommendations from health authorities on the consumption of cholesterol-rich foods and the upper limits for circulating cholesterol (approximately 2 g/L). In the absence of data on specific biochemical mechanisms involving cholesterol in brain functions, these standards remain favorable for the cardiovascular system. Maintaining a healthy vascular network, including that supplying the brain, should certainly participate in maintaining an optimal cognitive function.

References

Afzal, S., Bojesen, S.E., Nordestgaard, B.G. 2014. Reduced 25-hydroxyvitamin D and risk of Alzheimer's disease and vascular dementia. *Alzheimer's & Dementia* 10:296–302.

Allaire, J., Couture, P., Leclerc, M., et al. 2016. A randomized, crossover, head-to-head comparison of eicosapentaenoic acid and docosahexaenoic acid supplementation to reduce inflammation markers in men and women: The Comparing EPA to DHA (ComparED) Study. *Am. J. Clin. Nutr.* 104:280–7.

Allinquant, B., Clamagirand, C., Potier, M.C. 2014. Role of cholesterol metabolism in the pathogenesis of Alzheimer's disease. *Curr. Opin. Clin. Nutr. Metab. Care* 17:319–23.

Annweiler, C. 2016. Vitamin D in dementia prevention. *Ann. N. Y. Acad. Sci.* 1367:57–63.

Annweiler, C., Bartha, R., Karras, S.N., et al. 2015a. Vitamin D and white matter abnormalities in older adults: A quantitative volumetric analysis of brain MRI. *Exp. Gerontol.* 63:41–7.

Annweiler, C., Dursun, E., Féron, F., et al. 2015b. 'Vitamin D and cognition in older adults': Updated international recommendations. *J. Intern. Med.* 277:45–57.

Annweiler, C., Fantino, B., Gautier, J., et al. 2012a. Cognitive effects of vitamin D supplementation in older outpatients visiting a memory clinic: A prepost study. *J. Am. Geriatr. Soc.* 60:793–5.

Annweiler, C., Herrmann, F.R., Fantino, B., et al. 2012b. Effectiveness of the combination of memantine plus vitamin D on cognition in patients with Alzheimer disease: A pre-post pilot study. *Cogn. Behav. Neurol.* 25:121–7.

Annweiler, C., Montero-Odasso, M., Llewellyn, D.J., et al. 2013. Meta-analysis of memory and executive dysfunctions in relation to vitamin D. *J. Alzheimer's Dis.* 37:147–71.

Assmann, K.E., Touvier, M., Andreeva, V.A., et al. 2015. Midlife plasma vitamin D concentrations and performance in different cognitive domains assessed 13 years later. *Br. J. Nutr.* 113:1628–37.

Barberger-Gateau, P., Letenneur, L., Deschamps, V., et al. 2002. Fish, meat, and risk of dementia: Cohort study. *Brit. Med. J.* 325:932–3.

Barberger-Gateau, P., Raffaitin, C., Letenneur, L., et al. 2007. Dietary patterns and risk of dementia: The Three-City cohort study. *Neurology* 69:1921–30.

Barberger-Gateau, P., Samieri, C., Féart, C., et al. 2013. Acides gras oméga-3 et déclin cognitif: la controverse. *Cahiers Nutr. Diét.* 48:170–4.

Boespflug, E.L., McNamara, R.K., Eliassen, J.C., et al. 2016. Fish oil supplementation increases event-related posterior cingulate activation in older adults with subjective memory impairment. *J. Nutr. Health Aging* 20:161–9.

Bowman, G.L., Dodge, H.H., Mattek, N., et al. 2013. Plasma omega-3 PUFA and white mattermediated executive decline in older adults. *Front. Aging Neurosci.* 5:92.

Brouwer-Brolsma, E.M., van der Zwaluw, N.L., van Wijngaarden, J.P., et al. 2015. Higer serum 25-hydroxyvitamin D and lower plasma glucose are associated with larger gray matter volume but not with white matter or total brain volume in Dutch community-dwelling older adults. *J. Nutr.* 145:1817–23.

Burlot, M.A., Braudeau, J., Michaelsen-Preusse, K.M., et al. 2015. Cholesterol 24-hydroxylase defect is implicated in memory impairments associated with Alzheimer-like Tau pathology. *Hum. Mol. Genet.* 24(21):5965–76. DOI: 10.1093/hmg/ddv268.

Canevari, L., Clark, J.B. 2007. Alzheimer's disease and cholesterol: The fat connection. *Neurochem. Res.* 32:739–50.

Carlsson, C.M., Gleason, C.E., Hess, T.M., et al. 2008. Effects of simvastatin on cerebrospinal fluid biomarkers and cognition in middle-aged adults at risk for Alzheimer's disease. *J. Alzheimer's Dis.* 13:187–97.

Chew, E.Y., Clemons, T.E., Agrón, E., et al. 2015. Effect of omega-3 fatty acids, lutein/zeaxanthin, or other nutrient supplementation on cognitive function The AREDS2 randomized clinical trial. *JAMA* 314:791–801.

Cole, G.M., Frautschy, S.A. 2010. DHA may prevent age-related dementia. *J. Nutr.* 140:869–74.

Crichton, G.E., Elias, M.F., Davey, A., et al. 2014. Higher HDL cholesterol is associated with better cognitive function: The Maine-Syracuse study. *J. Int. Neuropsychol. Soc.* 20:961–70.

Crisby, M. 2006. The role of pleiotropic effects of statins in dementia. *Acta Neurol. Scand.* 185:115–18.

Cunnane, S.C., Plourde, M., Pifferi, F., et al. 2009. Fish, docosahexaenoic acid and Alzheimer's disease. *Prog. Lipid Res.* 48:239–56.

Curran-Celentano, J., Hammond, B.R., Ciulla, T.A., et al. 2001. Relation between dietary intake, serum concentrations, and retinal concentrations of lutein and zeaxanthin in adults in a Midwest population. *Am. J. Clin. Nutr.* 74:796–802.

Daiello, L.A., Gongvatana, A., Dunsiger, S., et al. 2015. Association of fish oil supplement use with preservation of brain volume and cognitive function. *Alzheimer Dement.* 11:226–35.

Dangour, A.D., Allen, E., Elbourne, D., et al. 2010. Effect of 2-y n-3 long-chain polyunsaturated fatty acid supplementation on cognitive function in older people: A randomized, double-blind, controlled trial. *Am. J. Clin. Nutr.* 91:1725–32.

Dangour, A.D., Andreeva, V.A., Sydenham, E., et al. 2012. Omega 3 fatty acids and cognitive health in older people. Brit. *J. Nutr.* 107:S152–8.

Danthiir, V., Hosking, D., Burns, N.R., et al. 2014. Cognitive performance in older adults is inversely associated with fish consumption but not erythrocyte membrane n-3 fatty acids. *J. Nutr.* 144:311–20.

D'Ascoli, T.A., Mursu, J., Voutilainen, S., et al. 2016. Association between serum long-chain omega-3 polyunsaturated fatty acids and cognitive performance in elderly men and women: The Kuopio Ischaemic Heart Disease Risk Factor Study. *Eur. J. Clin. Nutr.* 70:970–5.

Denis, I., Potier, B., Heberden, C., et al. 2015. Omega-3 polyunsaturated fatty acids and brain aging. *Curr. Opin. Clin. Nutr. Metab. Care* 18:139–46.

Devassy, J.G., Leng, S., Gabbs, M., et al. 2016. Omega-3 polyunsaturated fatty acids and oxylipins in neuroinflammation and management of Alzheimer's disease. *Adv. Nutr.* 7:905–16.

de Waal, H., Stam, C.J., Lansbergen, M.M., et al. 2014. The effect of souvenaid on functional brain network organisation in patients with mild Alzheimer's disease: A randomised controlled study. *PLoS One* 9:e86558.

Ding, Y., Qiao, A., Wang, Z., et al. 2008. Retinoic acid attenuates beta-amyloid deposition and rescues memory deficits in an Alzheimer's disease transgenic mouse model. *J. Neurosci.* 28:11622–34.

Devore, E.E., Grodstein, F., van Rooij, F.J., et al. 2010. Dietary antioxidants and long-term risk of dementia. *Arch. Neurol.* 67:819–25.

Dysken, M.W., Sano, M., Asthana, S., et al. 2014. Effect of vitamin E and memantine on functional decline in Alzheimer disease: The TEAM-AD VA cooperative randomized trial. *JAMA* 311:33–44.

Egert, S., Rimbach, G., Huebbe, P. 2012. ApoE genotype: From geographic distribution to function and responsiveness to dietary factors. *Proc. Nutr. Soc.* 71:410–24.

Eyles, D.W., Burne, T.H., McGrath, J.J. 2013. Vitamin D, effects on brain development, adult brain function and the links between low levels of vitamin D and neuropsychiatric disease. *Front. Neuroendocrinol.* 34:47–64.

Farina, N., Isaac, M.G., Clark, A.R., et al. 2012. Vitamin E for Alzheimer's dementia and mild cognitive impairment. *Cochrane Database Syst. Rev.* 11:CD002854.

Feeney, J., Finucane, C., Savva, G.M., et al. 2013. Low macular pigment optical density is associated with lower cognitive performance in a large, population-based sample of older adults. *Neurobiol. Aging* 34:2449–56.

Fukui, K., Ferris, H.A., Kahn, C.R. 2015. Effect of cholesterol reduction on receptor signaling in neurons. *J. Biol. Chem.* 290:26383–92.

Gamba, P., Testa, G., Gargiulo, S., et al. 2015. Oxidized cholesterol as the driving force behind the development of Alzheimer's disease. *Front. Aging. Neurosci.* 7:119.

Gil, R. 2010. *Neuropsychologie*. Paris: Masson.

Goodman, A.B., Pardee, A.B. 2003. Evidence for defective retinoid transport and function in late onset Alzheimer's disease. *Proc. Natl. Acad. Sci. U. S. A.* 100:2901–5.

Granholm, A.C., Bimonte-Nelson, H.A., Moore, A.B., et al. 2008. Effects of a saturated fat and high cholesterol diet on memory and hippocampal morphology in the middle-aged rat. *J. Alzheimer's Dis.* 14:133–45.

Grodstein, F., Chen, J., Willett, W.C. 2003. Highdose antioxidant supplements and cognitive function in community-dwelling elderly women. *Am. J. Clin. Nutr.* 77:975–84.

Gu, Y., Brickman, A.M., Stern, Y., et al. 2015. Mediterranean diet and brain structure in a multiethnic elderly cohort. *Neurology* 85:1744–51.

Heude, B., Ducimetière, P., Berr, C., et al. 2003. Cognitive decline and fatty acid composition of erythrocyte membranes – The EVA Study. *Am. J. Clin. Nutr.* 77:803–8.

Huang, T.L., Zandi, P.P., Tucker, K.L., et al. 2005. Benefits of fatty fish on dementia risk are stronger for those without APOE epsilon4. *Neurology* 65:1409–14.

Jernerén, F., Elshorbagy, A.K., Oulhaj, A., et al. 2015. Brain atrophy in cognitively impaired elderly: The importance of long-chain ω-3 fatty acids and B vitamin status in a randomized controlled trial. *Am. J. Clin. Nutr.* 102:215–21.

Johnson, E.J., McDonald, K., Caldarella, S.M., et al. 2008. Cognitive findings of an exploratory trial of docosahexaenoic acid and lutein supplementation in older women. *Nutr. Neurosci.* 11:75–83.

Johnson, E.J., Vishwanathan, R., Johnson, M.A., et al. 2013. Relationship between serum and brain carotenoids, α-tocopherol, and retinol concentrations and cognitive performance in the oldest old from the Georgia centenarian Study. *J. Aging Res.* 2013:951786.

Kalmijn, S., Feskens, E.J., Launer, L.J., et al. 1997. Polyunsaturated fatty acids, antioxidants, and cognitive function in very old men. *Am. J. Epidemiol.* 145:33–41.

Kang, J.H., Ascherio, A., Grodstein, F., et al. 2005. Fruit and vegetable consumption and cognitive decline in aging women. *Ann. Neurol.* 57:713–20.

Kesse-Guyot, E., Péneau, S., Ferry, M., et al. 2011. Thirteen-year prospective study between fish consumption, long-chain n-3 fatty acids intakes and cognitive function. *J. Nutr. Health Aging* 15:115–20.

Kim, D.H., Grodstein, F., Rosner, B., et al. 2013. Seafood types and age-related cognitive decline in the Women's Health Study. *J. Gerontol.* 68:1255–62.

Klapcinska, B., Derejczyk, J., WieczorowskaTobis, K., et al. 2000. Antioxidant defense in centenarians (a preliminary study). *Acta Biochim. Pol.* 47:281–92.

La Fata, G., Weber, P., Mohajeri, M.H. 2014. Effects of vitamin E on cognitive performance during ageing and in Alzheimer's disease. *Nutrients* 6:5453–72.

Latimer, C.S., Brewer, L.D., Searcy, J.L., et al. 2014. Vitamin D prevents cognitive decline and enhances hippocampal synaptic function in aging rats. *Proc. Natl. Acad. Sci. U. S. A.* 111:E4359–66.

Lee, L.K., Shahar, S., Chin, A.V., et al. 2013. Docosahexaenoic acid-concentrated fish oil supplementation in subjects with mild cognitive impairment (MCI): A 12-month randomised, double-blind, placebo-controlled trial. *Psychopharmacology* 225:605–12.

Léger, J.M., Mas, J.L. 2015. *Démences*. Paris: Doin.

Leray, C. 2015. *Lipids. Nutrition and health*. Boca Raton, FL: CRC Press.

Li, L., Cao, D., Garber, D.W., et al. 2003. Association of aortic atherosclerosis with cerebral beta-amyloidosis and learning deficits in a mouse model of Alzheimer's disease. *Am. J. Pathol.* 163:2155–64.

Littlejohns, T.J., Henley, W.E., Lang, I.A., et al. 2014. Vitamin D and the risk of dementia and Alzheimer disease. *Neurology* 83:920–8.

Liu, Y., Yu, J.T., Wang, H.F., et al. 2015. APOE genotype and neuroimaging markers of Alzheimer's disease: Systematic review and metaanalysis. *J. Neurol. Neurosurg. Psychiatry* 86:127–34.

Llewellyn, D.J., Langa, K.M., Lang, I.A. 2009. Serum 25-hydroxyvitamin D concentration and cognitive impairment. *J. Geriatr. Psychiatry Neurol.* 22:188–95.

Llewellyn, D.J., Lang, I.A., Langa, K.M., et al. 2010. Vitamin D and risk of cognitive decline in elderly persons. *Arch. Intern. Med.* 170:1135–41.

Lloret, A., Badía, M.C., Mora, N.J., et al. 2009. Vitamin E paradox in Alzheimer's disease: It does not prevent loss of cognition and may even be detrimental. *J. Alzheimer's Dis.* 17:143–9.

Loef, M., Walach, H. 2013. The omega-6/omega-3 ratio and dementia or cognitive decline: a systematic review on human studies and biological evidence. *J. Nutr. Gerontol. Geriatr.* 32:1–23.

Lopes da Silva, S., Vellas, B., Elemans, S., et al. 2014. Plasma nutrient status of patients with Alzheimer's disease: systematic review and metaanalysis. *Alzheimer's Dement.* 10:485–502.

Maddock, J., Cavadino, A., Power, C., et al. 2015. 25-hydroxyvitamin D, APOE 4 genotype and cognitive function: Findings from the 1958 British birth cohort. *Eur. J. Clin. Nutr.* 69:505–8.

Mangialasche, F., Solomon, A., Kåreholt, I., et al. 2013a. Serum levels of vitamin E forms and risk of cognitive impairment in a Finnish cohort of older adults. *Exp. Gerontol.* 48:1428–35.

Mangialasche, F., Westman, E., Kivipelto, M., et al. 2013b. Classification and prediction of clinical diagnosis of Alzheimer's disease based on MRI and plasma measures of α-/γ-tocotrienols and γ-tocopherol. *J. Intern. Med.* 273:602–21.

Mazereeuw, G., Lanctôt, K.L., Chau, S.A., et al. 2012. Effects of ω-3 fatty acids on cognitive performance: A meta-analysis. *Neurobiol. Aging* 33:1482.e17–29.

McGrath, J., Feron, F., Eyles, D., et al. 2001. Vitamin D: The neglected neurosteroid? *Trends Neurosci.* 24:570–1.

Michaelson, D.M. 2014. APOE ε4: The most prevalent yet understudied risk factor for Alzheimer's disease. *Alzheimer's Dement.* 10:861–8.

Miller, J.W., Harvey, D.J., Beckett, L.A., et al. 2015. Vitamin D status and rates of cognitive decline in a multiethnic cohort of older adults. *JAMA Neurol.* 72:1295–303.

Mingaud, F., Mormede, C., Etchamendy, N., et al. 2008. Retinoid hyposignaling contributes to aging-related decline in hippocampal function in short-term/working memory organization and long-term declarative memory encoding in mice. *J. Neurosci.* 28:279–91.

Minihane, A.M., Khan, S., Leigh-Firbank, E.C., et al. 2000. ApoE polymorphism and fish oil supplementation in subjects with an atherogenic lipoprotein phenotype. *Arterioscler. Thromb. Vasc. Biol.* 20:1990–7.

Misner, D.L., Jacobs, S., Shimizu, Y., et al. 2001. Vitamin A deprivation results in reversible loss of hippocampal long-term synaptic plasticity. *Proc. Natl. Acad. Sci. U. S. A.* 98:11714–19.

Moon, J.H., Lim, S., Han, J.W., et al. 2015. Serum 25-hydroxyvitamin D level and the risk of mild cognitive impairment and dementia: The Korean Longitudinal Study on Health and Aging (KloSHA). *Clin. Endocrinol.* 83:36–42.

Morell, P., Jurevics, H. 1996. Origin of cholesterol in myelin. *Neurochem. Res.* 21:463–70.

Morris, M.C., Brockman, J., Schneider, J.A., et al. 2016. Association of seafood consumption, brain mercury level, and APOEε4 status with brain neuropathology in older adults. *JAMA* 315:489–97.

Morris, M.C., Evans, D.A., Bienias, J.L., et al. 2002. Dietary intake of antioxidant nutrients and the risk of incident Alzheimer's disease in a biracial community study. *JAMA* 287:3230–7.

Morris, M.C., Schneider, J.A., Li, H., et al. 2015a. Brain tocopherols related to Alzheimer's disease neuropathology in humans. *Alzheimer's & Dementia* 11:32–9.

Morris, M.C., Tangney, C.C., Wang, Y., et al. 2015b. MIND diet associated with reduced incidence of Alzheimer's disease. *Alzheimer's Dementia* 11:1007–14.

Nolan, J.M., Loskutova, E., Howard, A.N., et al. 2014. Macular pigment, visual function, and macular disease among subjects with Alzheimer's disease: An exploratory study. *J. Alzheimers Dis.* 42:1191–202.

Nurk, E., Drevon, C.A., Refsum, H., et al. 2007. Cognitive performance among the elderly and dietary fish intake: The Hordaland Health Study. *Am. J. Clin. Nutr.* 86:1470–8.

Ono, K., Yamada, M. 2012. Vitamin A and Alzheimer's disease. *Geriatr. Gerontol.* 12:180–8.

Otsuka, R., Tange, C., Nishita, Y., et al. 2014. Serum docosahexaenoic and eicosapentaenoic acid and risk of cognitive decline over 10 years among elderly Japanese. *Eur. J. Clin. Nutr.* 68:503–9.

Patrick, R.P., Ames, B.N. 2015. Vitamin D and the omega-3 fatty acids control serotonin synthesis and action, part 2: Relevance for ADHD, bipolar disorder, schizophrenia, and impulsive behavior. *FASEB J.* 29:2207–22.

Peneau, S., Galan, P., Jeandel, C., et al. 2011. Fruit and vegetable intake and cognitive function in the SU.VI.MAX 2 prospective study. *Am. J. Clin. Nutr.* 94:1295–303.

Perkins, A.J., Hendrie, H.C., Callahan, C.M., et al. 1999. Association of antioxidants with memory in a multiethnic elderly sample using the third national health and nutrition examination survey. *Am. J. Epidemiol.* 150:37–44.

Pierrot, N., Lhommel, R., Quenon, L., et al. 2015. Targretin improves cognitive and biological markers in a patient with Alzheimer's disease. *J. Alzheimers Dis.* 49:271–6.

Poirier, J., Miron, J., Picard, C., et al. 2014. Apolipoprotein E and lipid homeostasis in the etiology and treatment of sporadic Alzheimer's disease. *Neurobiol. Aging* 35:3–10.

Przybelski, R., Agrawal, S., Krueger, D., et al. 2008. Rapid correction of low vitamin D status in nursing home residents. *Osteoporos Int.* 19:1621–8.

Qin, B., Plassman, B.L., Edwards, L.J., et al. 2014. Fish intake is associated with slower cognitive decline in Chinese older adults. *J. Nutr.* 144:1579–85.

Quinn, J.F., Raman, R., Thomas, R.G., et al. 2010. Docosahexaenoic acid supplementation and cognitive decline in Alzheimer disease: A randomized trial. *JAMA* 304:1903–11.

Raji, C.A., Erickson, K.I., Lopez, O.L., et al. 2014. Regular fish consumption and age-related brain gray matter loss. *Am. J. Prev. Med.* 47:444–51.

Ren, H., Luo, C., Feng, Y., et al. 2016. Omega-3 polyunsaturated fatty acids promote amyloid-β clearance from the brain through mediating the function of the glymphatic system. *FASEB J.* 31(1):282–93. DOI: 10.1096/fj.201600896.

Rondanelli, M., Opizzi, A., Faliva, M., et al. 2012. Effects of a diet integration with an oily emulsion of DHA-phospholipids containing melatonin and tryptophan in elderly patients suffering from mild cognitive impairment. *Nutr. Neurosci.* 5:46–54.

Ronnlund, M., Nyberg, L., Bäckman, L., et al. 2005. Stability, growth, and decline in adult life span development of declarative memory: Cross-sectional and longitudinal data from a population-based study. *Psychol. Aging* 20:3–18.

Rossom, R.C., Espeland, M.A., Manson, J.E., et al. 2012. Calcium and vitamin D supplementation and cognitive impairment in the women's health initiative. *J. Am. Geriatr. Soc.* 60(12):2197–205

Salem, N., Vandal, M., Calon, F. 2015. The benefit of docosahexaenoic acid for the adult brain in aging and dementia. *Prostaglandins Leukot. Essent. Fatty Acids* 92:15–22.
Scheltens, P., Twisk, J.W., Blesa, R., et al. 2014. Efficacy of Souvenaid in mild Alzheimer's disease: Results from a randomized, controlled trial. *J. Alzheimers Dis.* 31:225–36.
Shah, R.C., Kamphuis, P.J., Leurgans, S., et al. 2013. The S-Connect study: Results from a randomized, controlled trial of Souvenaid in mild-to-moderate Alzheimer's disease. *Alzheimer's Res. Ther.* 5:59.
Shen, L., Ji, H.F. 2015. Vitamin D deficiency is associated with increased risk of Alzheimer's disease and dementia: Evidence from meta-analysis. *Nutr. J.* 14:76.
Singh-Manoux, A., Kivimaki, M., Glymour, M.M., et al. 2012. Timing of onset of cognitive decline: Results from Whitehall II prospective cohort study. *BMJ.* 344:d7622.
Slinin, Y., Paudel, M., Taylor, B.C., et al. 2012. Association between serum 25(OH) vitamin D and the risk of cognitive decline in older women. *J. Gerontol. A Biol; Sci; Med. Sci.* 67:1092–8.
Small, B.J., Rosnick, C.B., Fratiglioni, L., et al. 2004. Apolipoprotein E and cognitive performance: A meta-analysis. *Psychol. Aging* 19:592–600.
Smith, M.A. 2006. Oxidative stress and iron imbalance in Alzheimer disease: How rust became the fuss! *J. Alzheimer's Dis.* 9:305–8.
Smith, A.D., Smith, S.M., de Jager, C.A., et al. 2010. Homocysteine-lowering by B vitamins slows the rate of accelerated brain atrophy in mild cognitive impairment: A randomized controlled trial. *PLoS One* 5:e12244.
Sodhi, R.K., Singh, N. 2014. Retinoids as potential targets for Alzheimer's disease. *Pharmacol. Biochem. Behav.* 120:117–23.
Solomon, A., Kåreholt, I., Ngandu, T., et al. 2007. Serum cholesterol changes after midlife and late-life cognition: Twenty-one-year followup study. *Neurology* 68:751–6.
Solomon, A., Kåreholt, I., Ngandu, T., et al. 2009. Serum total cholesterol, statins and cognition in non-demented elderly. *Neurobiol. Aging* 30:1006–9.
Sonnenberg, J., Luine, V.N., Krey, L.C., et al. 1986. 1,25-Dihydroxyvitamin D3 treatment results in increased choline acetyltransferase activity in specific brain nuclei. *Endocrinology* 118:1433–9.
Sparks, D.L., Scheff, S.W., Hunsaker, J.C., et al. 1994. Induction of Alzheimer-like beta-amyloid immunoreactivity in the brains of rabbits with dietary cholesterol. *Exp. Neurol.* 126:88–94.
Stoney, P.N., McCaffery, P. 2016. A vitamin on the mind: new discoveries on control of the brain by vitamin A. *World Rev. Nutr. Diet.* 115:98–108.
Street, S.J., Parletta, N., Milte, C., et al. 2015. Interaction of erythrocyte eicosapentaenoic acid and physical activity predicts reduced risk of mild cognitive impairment. *Aging Ment. Health* 19:885–91.
Swan, G.E., LaRue, A., Carmelli, D., et al. 1992. Decline in cognitive performance in aging twins. Heritability and biobehavioral predictors from the National Heart, Lung, and Blood Institute Twin Study. *JAMA Neurol.* 49:476–81.

Tan, Z.S., Harris, W.S., Beiser, A.S., et al. 2012. Red blood cell ω-3 fatty acid levels and markers of accelerated brain aging. *Neurology* 78:658–64.

Titova, O.E., Sjögren, P., Brooks, S.J., et al. 2013. Dietary intake of eicosapentaenoic and docosahexaenoic acids is linked to gray matter volume and cognitive function in elderly. *Age* 35:1495–505.

Tot Babberich Ede, N., Gourdeau, C., Pointel, S., et al. 2015. Biology of subjective cognitive complaint amongst geriatric patients: Vitamin D involvement. *Curr. Alzheimer Res.* 12:173–8.

Touyarot, K., Bonhomme, D., Roux, P., et al. 2013. A mid-life vitamin A supplementation prevents age-related spatial memmory deficits and hippocampal neurogenesis alterations through CRABP-I. *PLoS One* 8:e72101.

van de Rest, O., Spiro, A., Krall-Kaye, E., et al. 2009. Intakes of (n-3) fatty acids and fatty fish are not associated with cognitive performance and 6-year cognitive change in men participating in the Veterans Affairs Normative Aging Study. *J. Nutr.* 139:2329–36.

van de Rest, O., Wang, Y., barnes, L.L., et al. 2016. APOE ε4 and the associations of seafood and long-chain omega-3 fatty acids with cognitive decline. *Neurology* 86:2063–70.

van der Schaft, J., Koek, H.L., Dijkstra, E., et al. 2013. The association between vitamin D and cognition: A systematic review. *Ageing Res. Rev.* 12:1013–23.

Vanmierlo, T., Bogie, J.F., Mailleux, J., et al. 2015. Plant sterols: friend or foe in CNS disorders? *Prog. Lipid Res.* 58:26–39.

Vanmierlo, T., Popp, J., Kölsch, H., et al. 2011. The plant sterol brassicasterol as additional CSF biomarker in Alzheimer's disease. *Acta Psychiatr. Scand.* 124:184–92.

Vercambre, M.N., Boutron-Ruault, M.C., Ritchie, K., et al. 2009. Long-term association of food and nutrient intakes with cognitive and functional decline: A 13-year follow-up study of elderly French women. *Brit. J. Nutr.* 102:419–27.

Wanamaker, B.L., Kristopher, J., Swiger, K.J., et al. 2015. Cholesterol, statins, and dementia: What the cardiologist should know. *Clin. Cardiol.* 38:243–50.

Whalley, L.J., Deary, I.J., Starr, J.M., et al. 2008. n-3 Fatty acid erythrocyte membrane content, APOE varepsilon4, and cognitive variation: An observational follow-up study in late adulthood. *Am. J. Clin. Nutr.* 87:449–54.

Wimo, A., Prince, M. 2010. World Alzheimer Report 2010: The Global Economic Impact of Dementia. Alzheimer's Disease International. http://www.alz.co.uk/research/files/WorldAlzheimerReport2010.pdf.

Witte, A.V., Kerti, L., Hermannstädter, H.M., et al. 2014. Long-chain omega-3 fatty acids improve brain function and structure in older adults. *Cerebral Cortex* 24:3059–68.

Wu, S., Ding, Y., Wu, F., et al. 2015. Omega-3 fatty acids intake and risks of dementia and Alzheimer's disease: A meta-analysis. *Neurosci. Biobehav. Rev.* 48:1–9.

Xu, W., Tan, L., Wang, H.F., et al. 2015. Metaanalysis of modifiable risk factors for Alzheimer's disease. *J. Neurol. Neurosurg. Psychiatry* 86:1299–306.

Xue-shan, Z., Juan, P., Qi, W., et al. 2016. Imbalanced cholesterol metabolism in Alzheimer's disease. *Clin. Chim. Acta* 456:107–14.

Yaffe, K., Barrett-Connor, E., Lin, F., et al. 2002. Serum lipoprotein levels, statin use, and cognitive function in older women. *Arch. Neurol.* 59:378–84.

Yurko-Mauro, K., Alexander, D.D., Van Elswyk, M.E. 2015. Docosahexaenoic acid and adult memory: A systematic review and meta-analysis. *PLoS One* 10:e0120391.

Yurko-Mauro, K., McCarthy, D., Rom, D., et al. 2010. Beneficial effects of docosahexaenoic acid on cognition in age-related cognitive decline. *Alzheimer's Dement.* 6:456–64.

Zamroziewicz, M., Paul, E.J., Rubin, R.D., et al. 2015. Anterior cingulate cortex mediates the relationship between O3PUFAs and executive functions in APOE e4 carriers. *Front. Aging Neurosci.* 7:87.

chapter five

Other neurological diseases

Neurological disorders discussed in this chapter are considered as integral part of the nervous system diseases. They most often affect the brain, but they also the spinal cord and nerves. Besides Alzheimer's disease (Chapter 4), three other major neurological diseases show motor and sometimes mental aspects at different times in life: Parkinson's disease (Section 5.1), multiple sclerosis (Section 5.2), and epilepsy (Section 5.3).

These disorders affect directly nerve cells, their structure or their function, and are very diverse in their mode of expression, with some of them being associated with aging. Unlike other disorders, the origin of their outbreaks is increasingly well understood, thus explaining the development of new pharmacological treatments. As the influence of the environment is more and more taken into account, the progressive introduction of nutritional aspects may be exercised for therapeutic as well as preventive purposes.

5.1 Parkinson's disease

Parkinson's disease is a chronic neurodegenerative disorder that is slowly changing and that has a poorly understood origin. It reaches first an area of a few millimeters located at the base of the brain and rich in dopamine neurons (locus niger or substantia nigra), the latter being also abundantly supplied with vitamin D receptors. The death of these cells is associated with disturbances in all the neural networks associated with other brain areas (striatum, thalamus, subthalamic nucleus). These neurons are specialized in manufacturing dopamine, a neurotransmitter involved in the control of the body movements, mainly those that are automatic such as walking. Nonmotor symptoms are also observed; they are associated with various disorders such as sleep problems, smell loss (anosmia), cognitive troubles, poor balance, pain, and depression, all associated in neural networks located in nondopaminergic central structures.

Next to Alzheimer's disease, Parkinson's disease is the second most common neurodegenerative disease. Described in 1817 by Dr. James Parkinson, it was redefined and named for its discoverer by Dr. Jean-Martin Charcot in late nineteenth century. This pathology has benefited from an effective treatment only in 1967 with the introduction of L-DOPA (or levodopa), an intermediate substance in the catecholamine synthesis. Curiously, there is no simple definition of the disease.

Neurologists diagnose the disease by noting the presence of the following symptoms: walking difficulty; slowness of movement (bradykinesia or akinesia); unilateral resting tremor of the hand, foot, or both; and stiffness or poor muscle relaxation (hypertonia). Importantly, however, these cardinal symptoms are not always associated with the disease. A positive response to L-DOPA treatment confirms the diagnosis that could be more difficult to make by the existence of nontypical or very diverse signs, such as depression, pain, and fatigue.

Each patient is a particular subject with his or her own clinical symptoms, with the evolution of the disease being unique and depending on many factors. Some signs such as difficulty in speaking may appear after several years or remain insignificant.

Parkinson's disease mainly affects people older than 60 years (mean age 58 years); 10% of them have symptoms at less than 50 years old, but the disease process begins actually 5–10 years before the first symptoms appear. Men are slightly more affected than women (55 men for 45 women). Evidence suggests that in some cases the disease may be inherited. An estimated 15%–25% of people with Parkinson's have a known relative with the disease. The etiology of the majority of cases of the disease remains unknown, with multifactorial theories suggesting genetic and environmental interactions.

Globally, more than 6 million people are affected; the disease is diagnosed in more than 300,000 people each year. However, it is difficult to know exactly how many have developed the disease because many people think their symptoms are due to normal aging (Campenhausen et al. 2005). Furthermore, the figures are likely influenced by design differences in studies, such as diagnostic criteria and methods of case ascertainment.

In the United States, there are about 1 million people (0.37% of the population) with the disease, but 3 million people could be undiagnosed. It seems that the incidence is higher in whites than in blacks, Hispanics, or Asians, with the latter having the lowest rate. Using US Medicare data, it has been established that Parkinson's disease prevalence (per 100,000) was 2170 in white men, but 1036 in blacks and 1140 in Asians (Willis et al. 2010).

In China alone there are more than 1.7 million people with Parkinson's disease. The prevalence per 100,000 of population is the highest in Albania (800), Egypt (557), Canada (317), Israel (256), and Japan (193), but the lowest in Norway (102), Thailand (95), Sweden (76), and Poland (66) (Pringsheim et al. 2014). In France, about 100,000 people are affected, with 8000 new cases reported each year.

In all countries, given the aging population, the incidence of the disease can only progress. As such, it has been estimated that the number of individuals affected worldwide will double by 2030.

Among the tests used for screening the pathology, the Unified Parkinson Disease Rating Scale is the most used and best validated tool for measuring the overall status of patients with Parkinson's disease.

This test includes examinations in four main sections:

- Mental and behavioral status
- Activities in daily life
- Motricity examination
- Responses to treatment

5.1.1 ω-3 Fatty acids

As outlined in Chapter 1, the long-chain ω-3 fatty acids, mainly eicosapentaenoic acid (EPA) and docosahexaenoic acid (DHA), are major constituents of the nervous structures and are known to play an important role in cognitive function (Section 3.1). These components have also obvious neuroprotective properties in inhibiting programmed cell death (apoptosis) of neurons, but also in promoting neurogenesis and myelination. Importantly, these properties are particularly affected in the Parkinson's disease. In animals deficient in ω-3 fatty acids, there is a major disturbance of the dopaminergic signaling, also known to be altered in Parkinson's disease (Zimmer et al. 2002). Conversely, in animal models DHA supplementation has led to an attenuation of the symptoms related to the disease and been linked to physiological and molecular disorders (Agim and Cannon 2015). This is consistent with the discovery in rats of toxic mechanisms triggered by ω-3 fatty acid deficiency at the level of the substantia nigra (Cardoso et al. 2014). Thus, it is hoped that such research will be extended to humans.

Even if everyone agrees to recognize that the consumption of ω-3 fatty acid–rich foods is an important factor in reducing the risk of Parkinson's disease, the epidemiological studies thus far do not provide clear results. More research is needed to clarify the specific role of these lipids. In some patients, a few attempts have tried a supplementation with DHA and have shown less depression symptoms that are often associated with the disease, but no effect on motricity problems. At the initiation of treatment, there is a need to determine the disease status, the optimal dose, and the duration of treatment. Unfortunately, the conditions required for an efficient supplementation are still hypothetical.

The complexity of Parkinson's disease involving motor, and also behavioral and cognitive symptoms, makes the investigation very difficult and lengthy; therefore, a large scientific cooperation is required. As for Alzheimer's disease, the success is certainly conditioned by an early detection, so that tests are conducted before there is too much destruction of dopaminergic neurons.

While waiting for more results, it is recommended to maintain an intake of ω-3 fatty acids of marine origin, not exceeding the recommended intake that can occur with only the diet. Adopting a Mediterranean (or Cretan) diet represents currently the best way to reduce the risk of developing Parkinson's disease, as for other degenerative diseases. If consumption of marine products is too low, supplementation should be considered at an early age and followed throughout life.

5.1.2 Vitamin D

As early as 1997, the observation of a vitamin D deficiency in patients with Parkinson's disease was reported by Y. Sato in Japan and was confirmed much later in an European population (Evatt et al. 2008) and then in a Chinese population (Wang et al. 2016). This condition seems undisputed today as it is supported by all subsequent investigations carried out in several countries (Zhao et al. 2013).

It has not been yet determined whether vitamin D deficiency is at the origin or at the worsening of this debilitating disease, or conversely whether it results from it. However, it is evident that the disease is often accompanied by osteoporosis, a bone status causing frequent fractures and known to be directly related to a vitamin D deficiency. If one considers the data confirming largely that this vitamin is an important actor in harmonious brain development, the natural trend is to establish links between deficiency and onset of the neurological disease. In Finland, where sun exposure is reduced and vitamin D deficiencies are frequent, a large study conducted in 2010 by P. Knekt tends to confirm the relevance of the previous hypothesis (Knekt et al. 2010). Indeed, taking into account all other factors, the number of patients who became ill over a period of 30 years was three times higher in the group with the lowest vitamin level than in the group with the highest vitamin level. Meanwhile, cognitive performances of nondemented patients seem to be also affected by vitamin D levels as the scores estimated with neuropsychological tests, including those for depression, are even better when the vitamin levels are among the highest (Peterson et al. 2013).

As for multiple sclerosis (Section 5.2.2), it was observed that the prevalence of Parkinson's disease follows a geographic gradient. In the United States, that prevalence rises gradually when patients are studied from south to north (Lux and Kurtzke 1987). Unfortunately, there are few systematic studies in very remote countries; therefore, it is too early to link with certainty these observations to only a decrease in sun exposure, with the latter being responsible for a decrease of the cutaneous vitamin D synthesis.

In the near future, research in this area will accelerate after the discovery of a close association between the Parkinson's disease and a polymorphism of the gene controlling the synthesis of the vitamin D receptors. A genomic study published in 2011 has reinforced that working hypothesis (Butler et al. 2011). Given the close relationships between vitamin D and the capacity of dopamine synthesis by the brain, vitamin D is now considered a neurohormone efficiently protecting nerve cells in both human and animal models (Eyles et al. 2013).

5.1.2.1 *Intervention studies*

What about a possible treatment of Parkinson's disease with vitamin D? When investigators have studied the effect of vitamin D in experimental models, cells in culture, or animals, they always observed beneficial effects. For example, an increased dopamine content in the striatum and substantia nigra was observed after a vitamin D treatment in the brain of rats with an early-onset Parkinson's disease (Smith et al. 2006). In humans, any hope is now admissible because a team from Tokyo Medicine University in Shinjuku, Japan, has shown that the treatment of patients with 1200 international units (IU) (30 μg)/day of vitamin D_3 stabilized the motor symptoms after 1 year (Suzuki et al. 2013).

It is hoped that other teams will contribute to that specific therapeutic aspect of Parkinson's disease. Currently, three trials involving vitamin D in the treatment of the disease are reported on the official registry of the National Institutes of Health (http://www.clinicaltrials.gov). Certainly, much work is needed to clarify the role of vitamin D in the pathogenesis and progression of Parkinson's disease. Right now, sufficient data have been revealed by clinicians to participate in combating the disease when blood analysis indicates a vitamin deficiency. The low cost and the safety of the treatment favor this approach, even if all evidence has not been yet provided.

In conclusion, it seems appropriate to advise people who seem to develop Parkinson's disease to request a blood analysis to establish their vitamin D status. In case of insufficiency or deficiency, it is recommended that they increase first sun exposure to correct that condition. If impossible or difficult, they must absorb regularly a vitamin D supplement. At the least, this treatment will improve the bone condition of patients, even if the nervous system remains an uncertain target. This advice is especially aimed at patients in the symptomatic stage, also called the "honeymoon period," which lasts 3–8 years. Within the framework of a still hypothetical prevention of Parkinson's disease, as for osteoporosis or many other conditions (Leray 2015), it seems prudent to recommend to the elderly, regardless their lifestyle, a permanent and adequate vitamin D_3 *(cholecalciferol) intake. Supply may be done classically with 25–50 μg of vitamin (1000–2000 IU) per day or more conveniently with 100,000 IU every month or every 2 months. The general practitioner should be asked to prescribe periodically an analysis of the circulating vitamin* D_3 *level.*

5.1.3 Vitamin E

As with all neurodegenerative diseases (Section 4.2.4), oxidative stress has been implicated in Parkinson's disease. Thus, the use of vitamin E (Section 7.5) as for other powerful antioxidants has been mentioned as a possible therapy that may protect nerve cells (Sies et al. 1992). Vitamin E is abundantly present in plants, especially in oils and derived products from whole grains. It is the natural protection of lipids from oxidation, the most efficient that humans can find in the diet.

Many studies have attempted to demonstrate a protective effect of vitamin E in the context of Parkinson's disease. This motivation is mainly due to the high vulnerability to oxidative stress of the substantia nigra, a brain area rich in dopaminergic neurons (Kidd 2000). Admittedly, the results of epidemiological studies have suggested a protective effect of vitamin E against the disease, even with moderate doses and often without higher efficiency for higher doses (Etminam et al. 2005).

If we examine, for example, one of the largest studies carried out in The Netherlands (Rotterdam Study) on more than 5300 individuals aged 55–95 years, the consumption of about 10 mg/day of vitamin E is associated with about 40% lower risk of Parkinson's disease (de Rijk et al. 1997). Conversely, other studies were unable to detect any effect.

In Japan, comparing patients with subjects without any neurodegenerative disease has shown that for an intake of plant vitamin E greater than 7 mg/day, the risk to develop Parkinson's disease was two times lower than for lesser amounts of vitamin E. Curiously, women displayed a still three times lower risk.

So far, the scientific literature seems to assign a unanimous protective role to vitamin E, at least to the forms included in the commonly eaten plants.

5.1.3.1 Intervention studies

Only one study performed over 20 years has been devoted to the vitamin E supplementation, but it was not conclusive. Despite a daily treatment with 1.34 g of tocopherol for over 1 year, no beneficial effect could be detected in patients with a Parkinson's disease diagnosed less than 5 years ago (Shoulson 1993). Contrary to what the authors suggested at that time, it seems difficult to assign this negative result to the doses used in the study being too small. Thus, it seems more realistic to rely on an advanced disease state for which it is known that more than half of dopaminergic cells in the substantia nigra are already destroyed.

The majority of the work has mainly demonstrated that plant consumption seems to be the most protective measure, but the assignment of results only to vitamin E is probably too restrictive. Furthermore, several authors have pointed out that the observed effects may also be due to the presence of other antioxidant substances, or even to some substances still unidentified in the vegetables consumed. Results obtained in animals seem to show that consumption of sesame oil strengthens the vitamin E effects on slowing the neurodegenerative process.

It seems therefore advisable to adopt the conclusions of the previous work even temporarily in maintaining very early a consumption of vegetable products sufficient to supply the body with at least 12 mg/day of vitamin E, as recommended by the medical authorities (Leray 2015). A larger intake of about 50 mg/day as a pharmaceutical supplement is strongly recommended for the elderly.

5.2 Multiple sclerosis

Multiple sclerosis (MS) was described for the first time in 1868 by Jean-Martin Charcot, the founder of the modern neurology, at the Salpêtrière Hospital in Paris, France. MS is an inflammatory and autoimmune disease that affects the white matter in the central nervous system (brain, optic nerves, and spinal cord). It results from a disruption of the body's defense system in which antibodies (auto-antibodies) against the antigens of the same body are considered as foreigners (auto-antigens). The disease involves inflammation mechanisms causing an early loss of oligodendrocytes, the myelin-producing cells, and a depletion of myelin in many axonal fibers with formation of fibrous structures called MS plaques. These demyelination plaques cause secondarily motor, sensory, and cognitive impairments because nerve impulses are slowed or blocked, depending on the extent of demyelination. These disorders have an unpredictable progression and often lead to an irreversible disability, altering gradually the function of all nerve pathways. Motor disorders (decreased muscle strength), sensory disturbances (e.g., hot–cold sensation, touch, tingling), balance disorders, or visual disturbances are then observable. In 85% of cases, there is a succession of crises with neurological symptoms that disappear in two-thirds of cases without late-occurring effects. In MS lesions, it has been shown that remyelination occurs at the edge of the plaques, where oligodendrocytes are numerous and may even proliferate. These observations have allowed suggested a reorganization of the axonal membrane with appearance of new ion channels along the nerve fiber. Nevertheless, a decrease in the neuronal density has been demonstrated in certain brain areas (cortex, thalamus). The attack frequency varies from one individual to another (on average, an attack per year). The time sequence of these episodes of relapses characterizes the form of the disease called relapsing-remitting. This form is present in more than 85% of patients at the beginning of the disease. Over the long term, the remitting stage turns into a progressively chronic phase wherein the neurological discomfort settles over several months without superimposed attack. Sometimes, this discomfort settles regularly and is called the progressive primary form of MS. This diversity of forms, originating probably from a diversity of causes, justifies the difficulty to develop treatments and the diversity of results obtained with environmental approaches (e.g., diet, sun exposure, vitamin intake). Besides neurological phenomena, MS displays very frequently cognitive disorders that are mainly related to memory deficits, spatial perception, reasoning, or attention. Unfortunately, these disorders have not been taken into account in clinical studies related to food and environment.

Pharmacological treatments currently used reduce the inflammatory attacks, but they cannot stop the progression of the disease. New therapeutic strategies slowing immune attacks and developing an onset of myelination are being evaluated.

Worldwide, there are about 2.3 million people with MS, including 300,000 in Western Europe, although the number may be much higher because it is likely that many ill people remain undiagnosed in certain parts of the world. Likewise, estimated annual incidence rates ranged widely from less than 1 to more than 10 in 100,000.

There has not been a scientifically sound national estimation of the prevalence of MS in the United States since 1975. Currently, the incidence or prevalence of this disease is not consistently reported. The US Multiple Sclerosis Foundation relies on estimates based on the earlier studies. It has been estimated that more than 400,000 people have the disease and that the average person has about a 1 in 750 chance (0.1%) of developing it. It is noticeable that the rates of the disease are higher farther from the equator: about 57–78 cases per 100,000 people in southern states (below the 37th parallel) and about 110–140 cases per 100,000 in northern states (above the 37th parallel). A very high prevalence was observed in Canada (290 per 100,000)

In France, there are more than 80,000 patients. Each year, from 3000 to 5000 new cases are diagnosed in young people, two-thirds of which are women. Nearly 70% of patients present the first symptoms between 20 and 50 years of age. MS is in the young the leading cause of acquired neurological disability and is the leading cause of disability in young adults after road accidents (French League against multiple sclerosis).

Genetic factors are thought to play a significant role in determining who develops MS. The twin of someone with MS has a 25% chance of developing the disease. The fact that the risk is only 1 in 4 demonstrates that other environmental factors, including geography, ethnicity, or nutrition are likely involved as well. Early on, diet was suspected as a possible way to intervene in the development of the disease, with the composition of dietary ω-3 fatty acids being considered a priority

(Section 5.2.1). The importance of sun exposure is another widely studied factor; it has been the subject of many investigations because it seems linked to statistically proven cyclic variations in the disease prevalence indicating clearly the existence of relationships with vitamin D (Section 5.2.2).

5.2.1 ω-3 Fatty acids

In 1956, H. Sinclair, the discoverer of the concept of essential fatty acids, expressed the assumption that MS could result from diets deficient in highly unsaturated fatty acids, such as ω-6 or ω-3 fatty acids. Because a ω-6 fatty acid deficiency is unlikely in Western countries, given the high consumption of vegetable oils rich in linoleic acid (18:2 ω-6) and of meat rich in arachidonic acid (20:4 ω-6), the attention focused naturally on ω-3 fatty acids. Later, analysis performed on postmortem brain specimens from patients with MS showed ω-3 fatty acid levels decreased by 50% compared to controls (Kishimoto et al. 1967). Already 50 years ago in the United States, Dr. R. L. Swank recommended for a year a diet low in fat, but rich in fatty acids derived from fish and plants to reduce the frequency and severity of attacks observed in his patients with MS. He had the idea to recommend a diet low in saturated fats (less than 20 g/day), but rich in vegetables and fish and fortified with 5 g/day of fish oil. He observed that during the first 7 years, the frequency and inflammatory severity of the attacks were reduced. He made a fundamental observation: the earlier patients adopted this diet, the better were the results. Moreover, Dr. Swank reported that after 4 years, only 8% of subjects had a worse situation if they had followed very early the diet against 65% among those who had started too late. Interestingly, a reactivation of the disease was observed in almost all cases if the recommended diet was stopped even after 5–10 years. After these results, Dr. Swank wrote his recommendations in the form of recipes in the successful book *The Multiple Sclerosis Diet Book*, originally published in 1972.

Some authors have used the geographical distribution of the disease to reinforce the attractive hypothesis of a therapeutic effect of the lipid constituents of fish oil. Thus, the lower incidence of the disease in Japan or China compared to Europe or the United States could be a consequence of higher fish consumption in the first two countries. The same difference was found between the inhabitants of the coastal zones of Norway and the rest of the population (Larsen et al. 1985). It is however not certain that the dietary intake of ω-3 fatty acids is the only explanation, because fish are also a vitamin D–rich source (Section 7.4). Moreover, a combined effect of both lipid types may not be excluded.

Several authors observed that fish consumption also had favorable effects on disease progression. For example, in Belgium, a study demonstrated that in patients with MS, an important consumption of fish decreased the attack severity, but it did not influence the slow evolution of the disease between attacks (D'hooghe et al. 2012). These results prove once again that the evaluation of the influence of the nutritional treatments on the disease remains complex.

Despite sporadic studies supporting a beneficial effect of ω-3 fatty acids, one of the largest clinical and nutritional investigations, done in Boston, Massachusetts, on nearly 200,000 women followed for about 14 years, was unable to detect any effect of these fatty acids and other lipids on the incidence and progression of the disease (Zhang et al. 2000). In contrast, an Australian survey of a 500-patient cohort showed that subjects consuming the highest amount of fish had an improved quality of life, a lower risk of attack, and less intense nervous disorders (Jelinek et al. 2013).

5.2.1.1 *Intervention studies*

Considerable research has nevertheless shown that ω-3 fatty acids protected nervous tissue and were able to modulate the immune responses, especially in animal models of the disease. Thus, it is understandable that clinicians have attempted to verify whether the intake of these fatty acids, in the form of dietary supplements, could favorably influence the evolution of MS. Since the 1970s, many studies have been carried out, but they have generally produced few convincing outcomes. Among the encouraging results is the research by I. Nordvik at the University of Bergen in Norway on patients suffering from MS and who were newly diagnosed and immediately subjected to a treatment with 0.9 g of ω-3 fatty acids per day (Nordvik et al. 2000). After a 2-year follow-up, the attack frequency was significantly reduced and the neurological condition improved compared to the situation before treatment. It is regrettable that the size of the studied group (16 patients) was too small to draw meaningful conclusions.

Another attractive study was done by Dr. B. Weinstock-Guttman at Buffalo General Hospital in New York, with patients whose diagnosis occurred during the last 3 years and who were supplemented daily with 6 g of fish oil for a year (Weinstock-Guttman et al. 2005). The authors recorded moderate beneficial effects on clinical manifestations of the disease and the quality of life. Notably, patients were instructed to have a diet with a lipid content lower than 15% of their total energy intake. The previously reported assumptions made by Dr. R. L. Swank are still found verified here. Despite positive results, the authors emphasized the significant weight losses in supplemented subjects. Other nutritional conditions should be used in future investigations to optimize these early results.

The latest and most interesting progress in the treatment of MS with fatty acids is the clinical work of Dr. M. C. Pantzaris at the neurological clinic in Nicosia, Cyprus (Pantzaris et al. 2013). This doctor tested in patients with a carefully characterized disease the combined effect of a daily intake of ω-3 fatty acids (6.3 g of EPA plus DHA) combined with a vitamin E component (760 mg of γ-tocopherol) compared to that of a placebo (olive oil). The major observation of this work was a significant 62% reduction in relapses after 2 years and a lower progression of the disease. The combination of ω-3 fatty acids and γ-tocopherol was the most successful compared to fatty acids (18%) or γ-tocopherol alone (30%). These results obtained in the absence of any pharmaceutical treatment highlight the importance of a long-term administration of a mixture of ω-3 fatty acids and a powerful antioxidant (γ-tocopherol). Many similar trials are required to optimize an MS treatment based on dietary lipids. It also is necessary to determine the efficiency of these supplementations in the presence of recent treatments, such as those involving monoclonal antibodies.

The analysis of positive results should not obscure studies that have not allowed conclusions usable for therapy, even in using larger groups of patients and similar supplementations. It seems inevitable that future research should be designed using more patients who are clinically well characterized and with better defined fatty acid mixtures. It is evident that long-term research hold back the initiatives for such investigations, but it would be desirable, in the interest of patients, that national support is promptly assigned to teams already engaged in this research.

In conclusion, pending more accurate results, it may only be recommended, as for other lipids, to maintain a dietary intake of ω-3 fatty acids of marine origin in accordance with the recommendations of the medical authorities (Section 7.2). The results reported above should only suggest to patients recently diagnosed with MS to adopt a level of dietary ω-3 fatty acids significantly higher (Two or three times) than that recommended for healthy people.

5.2.2 Vitamin D

The influence of vitamin D on the immune system has been studied for more than 25 years; a comprehensive review of particular physiological aspects is given by Schoindre et al. (2012). The evolution of our knowledge is based on many epidemiological investigations linking vitamin D and several autoimmune diseases and has also resulted from the demonstration of a vitamin D synthesis in immune cells.

In 1974, after some preliminary observations, Dr. P. Goldberg hypothesized that subjects genetically predisposed to MS should have a vitamin D supply higher than that of healthy subjects (Goldberg 1974). A vitamin D deficiency during adolescence could thus lead to the formation of an abnormal myelin or even to demyelination plaques followed by the neurological symptoms specific of the disease. Therefore, a vitamin D supplementation may well induce its beneficial effects in young adults where the myelin formation is near completion. Goldberg applied his theory in treating young subjects (22–37 years) with cod liver oil (20 g/day) (Goldberg et al. 1986). The very positive results (2.4 times less relapses) after 2 years with an intake corresponding to about 5000 IU of vitamin D per day led Goldberg to put the benefit on the account of the vitamin D contained in cod liver oil. Cautiously, he also spoke of a possible action of fish oil fatty acids. Although valuable in the fight against the clinical manifestations of the disease, these findings highlight the difficulty of this type of nutritional study, especially when they involve complex natural products. The fact remains that fish oil, whatever the responsible component, seems to have a significant efficiency in limiting the progression of MS.

Since then and in many countries, several studies have confirmed that vitamin D deficiency is one risk factor in the development of MS. Indeed, as for other pathologies, one of the characteristics of the disease is its geographical distribution, with a greater frequency in low sunshine areas where the skin vitamin D synthesis is reduced and without any nutritional influence.

As for the Parkinson's disease (Section 5.1.2), the geographical distribution of MS is not uniform worldwide: it is rare in equatorial regions, with its incidence increasing gradually north from the equator. For example, there are high-prevalence areas (100–300 per 100,000) in Scandinavia, Scotland, northern Europe, Canada, and the northern United States; average-prevalence areas (more than 30) in Europe, Russia, and Australia; moderate-prevalence areas (5–30) around the Mediterranean basin, south of the United States, and Africa; and low-prevalence areas (less than 5) in East Asia, India, Africa, the Caribbean, and Mexico. However, geographical distribution depending on the latitude cannot be generalized to the entire world population. Numerous exceptions show that other factors also are involved, with ethnicity being the most important. Thus, exceptions to the latitudinal gradient in the Italian region and northern Scandinavia are likely a result of genetic and behavioral-cultural variations (Simpson et al. 2011).

The relationship between disease prevalence and sunshine has been well studied in North America (Kurtzke et al. 1979) and France. In France, Dr. S. Vukusic observed, by using the statistics established by the insurance of farmers (Mutualité Sociale Agricole), that the distribution of patients was heterogeneous, but that it was correlated with latitude. Indeed, the prevalence of the disease was twice as high among farmers living in the northeastern areas (about 100 per 100,000 inhabitants) compared to those living in the southwestern regions (about 50 per 100,000) (Vukusic et al. 2007).

Accurate surveys have even shown that subjects going out for short times in the sun, either in summer in countries with little sunshine (Norway) or year-round in sunny countries (Mexico, Italy), have a high risk

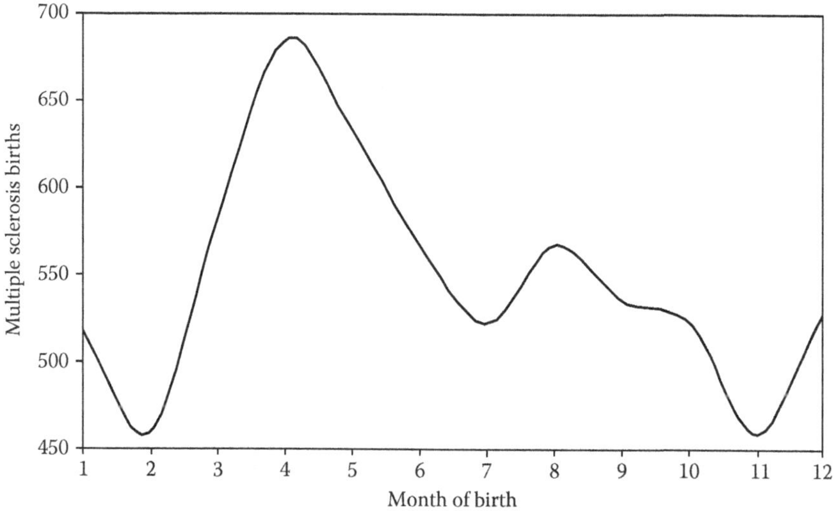

Figure 5.1 Frequency of the number of births of Norwegians suffering from multiple sclerosis and born from 1930 to 1979 based on the month of the year. (Modified from Torkildsen, O., et al., *Ann. Clin. Transl. Neurol.*, 1, 141–144, 2014.)

of suffering from MS. Other risks include the frequent use of sunscreens in the young, the very fair skinned, and those with red or blonde hair.

After counting statistics for nearly 80,000 English, an even more amazing relationship has been demonstrated (Dobson et al. 2013). This work described that there was a greater risk of developing MS in adults when the mother's pregnancy took place during the autumn and winter, periods of low sun exposure. A review of 15 studies published on this topic and done in different countries confirmed that the month of birth influenced distinctly the risk of developing the disease. As for schizophrenia (Section 6.1.3.2), the effect seems amplified in less sunny areas (Figure 5.1) (Torkildsen et al. 2014), but remains detectable as well in very sunny countries such as Kuwait (Akhtar et al. 2015).

It has been repeatedly shown that the presence of MS is associated with low vitamin D levels, with the deficiency resulting either from reduced exposure to sunshine, an inadequate food supply, or both. Almost all the scientific literature agrees on this association. One can find one confirmation in a review made from 11 studies that highlighted the general consensus that patients with MS have consistently low blood vitamin D levels (Duan et al. 2014). Dr. E. Thouvenot in the Nîmes Hospital, France, also showed that the level of the vitamin D deficiency was well correlated with the degree of patient disability, thus upholding the interest for a supplementation to slow the disease progression (Thouvenot et al. 2015). A study of patients living in Poland

and receiving immune-modulating drugs has shown that hypovitaminosis D is more prevalent during winter than summer, both in patient group and the controls, especially in female patients with higher levels of disability. Furthermore, low vitamin D levels were associated with a more severe course of disease and an increased number of relapses (Brola et al. 2016).

To accurately investigate the possible links between the risk of developing disease and vitamin D deficiency, it is necessary to examine a large population considered as healthy and to study it several years later when some individuals have developed the disease. Such a study was carried out by the team of Prof. A. Ascherio at Harvard Medical School, Boston, Massachusetts (Munger et al. 2006). These researchers followed hundreds of young soldiers recruited between 1993 and 2004. None of them was suffering from neurological disease in recruitment before the age of 20 years, all having a blood vitamin D analysis. After a little over 4 years and several blood explorations, 257 cases of MS appeared. The authors of that study observed that while comparing these subjects to healthy subjects, the risk of MS significantly decreased with increasing levels of vitamin D. However, the relationship was less clear among the African American subjects, probably in connection with their very low initial vitamin D level. Some years later, Dr. Ascherio established that about three quarters of MS cases could be prevented if children and young adults were maintaining a blood vitamin D level greater than 40 ng/mL (100 nmol/L).

In the past 5 years, several publications on that topic have focused on the statistical evidence of a close correlation between vitamin D and MS. Thus, Dr. Ascherio clearly showed that in patients treated with interferon β-1b, the highest levels of vitamin D allowed to predict a reduced disease activity and also the lowest progression rates (Ascherio et al. 2014). These findings are more and more frequently noticed, although the nature of the mechanisms involved are not fully elucidated.

A recent discovery has specified a missing link in these complex connections between vitamin D and MS. Indeed, the study carried out by the team of Dr. B. Richards at McGill University, Montreal, Quebec, focused on the characterization of low vitamin D levels in subjects with specific nucleotide polymorphisms and the likelihood of developing MS. That study was conducted with 14,498 subjects with the disease and 24,091 healthy subjects (Mokry et al. 2015). It seemed that from birth, children with genetically lower vitamin D levels had a risk two times higher for developing the disease, with the latter being diagnosed 20–50 years later.

5.2.2.1 *Intervention studies*

It is too early to assert that supplementing healthy children and adults with vitamin D will reduce the risk of developing MS much later; the clinical trials undertaken for the past 10 years remain still unconvincing.

A recent review of all the clinical studies in this area has highlighted the ambiguity of the results.

What are the causes of these failures when the previously exposed data suggested that we are getting closer to the goal? No serious leads are currently considered. Given the diversity of protocols, it is possible that the doses of vitamin D and the duration of treatment were insufficient. Moreover, the period of the start of treatment may be inappropriate. It is certainly desirable that the prevention starts very early, in early childhood, and long before the first symptoms of the disease. In the subjects genetically deficient in vitamin D, the results also suggested the interest of preventive measures, adapted according to the genome or the examination of the vitamin status.

Several large clinical trials are studying whether adding vitamin D to the treatment of patients with interferon may offer a therapeutic advantage. In France, another way is being explored in the Programme Hospitalier de Recherche Clinique (PHRC). Dr. Eric Thouvenot at University of Montpellier, thinks that the effect of interferon will mask that of vitamin D during the disease treatment. So, he attempted to check whether vitamin D administered when the first symptoms are reported could reduce the risks that a disease not yet confirmed could be converted into MS. Initiated in 2013, that study is being conducted in 33 centers across the country; the results will be known likely by 2018.

The French Society of Multiple Sclerosis ignores the research advances on vitamin D, as does the Multiple Sclerosis Association of America, and the US National Multiple Sclerosis Society. However, the Multiple Sclerosis Society of Canada informs its members, gives some dietary advice, and states that many doctors already prescribe a daily intake of 1000–2000 IU of vitamin D. In addition, the society website said that any dietary modification requires advice from a doctor or a nutritionist. The society funds several research teams on that subject and gives its members ample information from published work worldwide (https://mssociety.ca/). It clearly states that "clinicians who treat people with multiple sclerosis should become familiar with current vitamin D information so that they are able to educate their patients, and, if appropriate, diagnose and treat vitamin D deficiency. Also, due to the inherited risk of multiple sclerosis and the possible preventative effect of vitamin D supplementation, it may be reasonable for clinicians to discuss the possible implications of vitamin D deficiency and supplementation for the children of parents with multiple sclerosis." (http://www.nationalmssociety.org/). This approach is in progress in several countries because many neurologists are showing interest in the supplementation of adolescents and young adults to prevent many cases of MS. In case of favorable results, it would be possible in the general interest to consider a monitoring of vitamin D status in the whole population.

In this area, studies are underway, and curiously more than for many other diseases of the nervous system. Indeed, in 2016, 37 clinical trials involving multiple sclerosis and vitamin D were listed in the National Institutes of Health database (https://clinicaltrials.gov), with seven being still at the stage of patient recruitment. Two major studies, including one in France, (CHOLINE study) just ended. Their aim was to determine whether adding vitamin D to the treatment of patients with interferon-β may have a therapeutic benefit. Thus, it is possible to expect new results in the near future on both prevention and treatment of this serious disease.

All studies discussed above indicate that the overall level of evidence for the involvement of vitamin D in the triggering and development of MS may be already considered high. From the perspective of prevention, it seems thus capital to detect any vitamin D deficiency in childhood and if necessary to begin a supplementation to restore normal circulating levels of this vitamin. As highlighted in 2011 by Dr. C. Pierrot-Deseilligny, working on MS in Salpêtrière Hospital in Paris, France, "rather than wait for the results of phase III trials that still will take several years, it seems wise from a preventive point of view to supplement patients with vitamin D, especially if they are deficient and have factors tending to worsen that condition."

5.3 *Epilepsy*

Epilepsy is a neurological disease known since ancient times and that remains, despite advances in medicine, the most common neurological pathology in the world irrespective of age, race, country, or geographic area. The word "epilepsy" comes from "epilepsia" (επτλ | πσ{α), which means attack. The disease is complex and multifactorial, involving a genetic background associated with environmental factors.

The troubles have various causes, mechanisms and events, leading sometimes to talk about epilepsy syndrome. The disease occurs in many ways: by recurrent and spontaneous seizures known as partial seizures resulting from brain injury, or as primary generalized seizures involving more widely distributed mechanisms in the brain. Seizures vary in frequency of less than once per year to several times per day. The primary generalized seizures are motor seizures, often impressive, shaking the limb muscles and they may include a loss of consciousness. This phase is sometimes accompanied with convulsive movements that could cause the fall of the subject and serious injury.

The seizure is caused by a disruption of the electrical communication between thousands of neurons becoming hyperexcitable and leading to an intense electrical discharge in the nervous networks. The wide variety of the disease symptoms derives from the various brain areas affected by the electric discharges and their propagation. Thus, when the discharge starts at the level of the motor cortex, stiffening or rhythmic jerks occur in different body areas. Other brain areas could be the anatomical origin of various types of hallucinations (sounds, visions, odors), memory reminiscences, or incomprehensible words.

The plurality of manifestations of epilepsy suggests the existence of a large variety of causes. Indeed, in many countries clinical studies have shown that epilepsy usually accompanied several cerebrovascular diseases or disorders such as ischemic stroke, trauma, carcinomas, and congenital disorders. It can be also present during neurodegenerative troubles, such as Alzheimer's disease and vascular dementia. If no cause has been diagnosed, it is then issued as idiopathic epilepsy, being mainly present in children and adolescents.

It has been estimated that 3% of the world population suffers from epilepsy seizures at some moment or another in life.

The average incidence of epilepsy each year in the United States is estimated at 150,000 or 48 for every 100,000 people. The number of people with epilepsy (prevalence) ranges from 1.3 million to 2.8 million (or 5 to 8.4 for every 1000 people), but a more accurate estimation is 2.2 million people or 7.1 for every 1000 people (US Epilepsy Foundation, http://www.epilepsy.com/learn/epilepsy-statistics).

In Europe, the number of patients with epilepsy has been estimated at 3 million (prevalence of 600 per 100,000 people) (Forsgren et al. 2005). The World Health Organization has determined that a European doctor has 10–20 people with epilepsy in his patient base. The total cost incurred by the disease exceeds 20 billion euros per year for the European continent.

In France, it has been estimated that nearly 400,000 people are suffering from seizures, with half of them being younger than 20 years, and that there are about 20,000 new cases per year. With the aging population, an increase in the proportion of patients can be expected, mainly because of cerebrovascular and neurodegenerative diseases.

The mortality rate of patients is higher than for the general population, but 8%–17% of deaths occur suddenly and remain unexplained. Several studies have linked this higher mortality to increased heart attacks during winter and to arrhythmia crises during and between seizures.

Current epilepsy treatments are very specific and aim to block the alterations in the transmission of nerve impulses from one neuron to another. The drugs act primarily by blocking synaptic ion channels at the level of synapses. Unfortunately, nearly 20% of patients do not respond favorably to these drugs (drug-resistant epilepsy) and may even exhibit a mortality rate 10 times higher than for the normal population. New treatments are just appearing, with globally better efficiency but inducing several more or less important side effects.

As for MS, two types of lipids have been investigated for prevention or treatment of epilepsy through a supplementation of ω-3 fatty acids (Section 5.3.1) or vitamin D (Section 5.3.2).

5.3.1 ω-3 Fatty acids

The ω-3 fatty acids (Section 7.1) have quickly attracted the interest of researchers wishing to limit the magnitude of seizures. Previously, it has been shown that these fatty acids were able to stabilize plasma membranes and partially inhibit electrical currents related to sodium and calcium movements in cardiac cells in culture (Xiao et al. 1995). This observation may explain the well-known antiarrhythmic effect of ω-3 fatty acids on the heart and could thus be meaningful for the exploration of the causes of seizures involving cerebral hyperexcitability. This heart effect was moreover described in animals fed with diets enriched with fish oil and in humans in some clinical studies after EPA and DHA supplementation (Leray 2015). It is remarkable that some drugs used as anticonvulsant against seizures are also able to inhibit the activity of voltage-dependent sodium channels. Phenytoin, one of the most prescribed antiepileptic drugs worldwide, sold under the brand name Dilantin, among others, is also used as both an antiepileptic and antiarrhythmic.

Except for some short-term experiments, the administration of ω-3 fatty acids, including linolenic acid, induced usually in animals beneficial effects on the manifestations of a pharmacological epilepsy (Taha et al. 2010).

5.3.1.1 Intervention studies

In humans, the first experimental study was published in 2002, at the Weizmann Institute in Rehovot, Israel. Unfortunately, it was realized with a small number of subjects (Schlanger et al. 2002). Five patients suffering generalized seizures were hospitalized and treated with several anticonvulsant products. Meanwhile, they were ingesting 3.25 g of ω-3 fatty acids daily (46% DHA, 18% EPA, and 1% linolenic acid) as a spreadable paste. After a 6-month treatment, the five patients underwent a marked decrease in the frequency of epilepsy seizures.

Several other published clinical studies have unfortunately provided only discordant results. These discrepancies may arise from treatment periods being too short (3 months) or from poorly adapted fatty acid doses, but also from different forms of the disease. The trial published in 2015 by Prof. C. M. DeGiorgio of the University of California–Los Angeles represents perhaps for the first time a significant hope for a long-term therapy in patients suffering from a form of epilepsy resistant to conventional drugs (DeGiorgio et al. 2015). It follows from that publication that a moderate dose of fish oil (about 1 g of ω-3 fatty acids per day administered for 10 weeks) reduced seizure frequency by approximately 34% compared with placebo (sunflower oil), with a two times dose of fish oil producing no effect. This research showed the importance of experimental design and deserves to be extended to a larger population of patients selected according to very specific clinical criteria. As Prof. DeGiorgio pointed out the treatment is efficient, safe, and inexpensive, while improving simultaneously the patients' cardiovascular system status. Even more significant results have been obtained in epileptic children refractory to traditional treatments and treated daily for only 3 months with 1.2 g of fish oil (240 mg of DHA and 360 mg of EPA) (Reda et al. 2015).

Based on these outcomes, it would be desirable that neurologists take into account that a possibility of treatment exists and with more data treatment can be offered to all patients suffering from epilepsy. It is hoped that other studies on wider and more diverse populations in various countries will confirm these encouraging results; in the United States, more than 2 million people could benefit and in France, nearly 100,000.

A strong argument for the use of dietary ω-3 fatty acids is their possible contribution to the antiepileptic effect of the ketogenic diet (or Atkins diet). Although difficult to tolerate, this diet, rich in lipid but with low in carbohydrate content, has been successfully used in the fight against seizures, especially those particularly resistant to drug treatments. Curiously, it has been shown that the ketogenic diet increased the blood content of ω-3 fatty acids at the expense of adipose tissue and liver, probably for a subsequent incorporation into the brain (Taha et al. 2005). That mechanism may even increase the effectiveness of the treatment and reduce the complications associated with the diet (Dahlin et al. 2007).

The work undertaken by Prof. C. M. DeGiorgio focused on the situation of patients with epilepsy from refractory to current treatments (DeGiorgio et al. 2015). Indeed, if these patients have two to three times more risk of premature death than the general population, the situation of 20%–30% of patients that are refractory to treatment is even more dramatic. The latter are very often victims of sudden death (up to 17% of deaths), an unpredictable outcome but likely connected to cardiac abnormalities (Stollberger and Finsterer 2004). Because ω-3 fatty acids are effective in the fight against many cardiovascular diseases, it is not surprising that they may also play a role in the protection of patients against cardiac complications that accompany epilepsy.

Experiments in animals have already produced encouraging results (Scorza et al. 2008), and many clinicians hope that patients with epilepsy refractory to drugs will soon get an easy treatment based on a moderate and inexpensive intake of natural ω-3 fatty acids. Is it possible to find international financial support to test this hypothesis on a very large number of people for many years? More than 3 million patients in Europe and 2 million in the United States are expecting these advances. The challenge seems so huge that the Brazilian specialist of these trials, Dr. R. A. Cysneiros, has alerted researchers worldwide about a possible lack of edible fish as a result of increasing pollution and climate change.

The limited success of past research and especially the complexity of protocols in this area probably explain the small number of studies in progress. Fortunately, height clinical trials are declared completed or recruiting in the international database (www.clinicaltrials.com), including an ongoing trial with phosphatidylserine-enriched in ω-3 fatty acids in the Hospices Civils in Lyon, France (Dr. S. Rheims).

In conclusion, clinicians and parents should agree with the epilepsy specialist Prof. Fulvio A. Scorza, Sao Paulo University, Brazil, that it is reasonable to ensure that young children, even healthy, maintain their ω-3 fatty acid intake consistent with the latest recommendations (Section 7.2). He also highlights that an abundant literature has already shown that eating fish or ingesting oil capsules enriched with ω-3 fatty acids is not only safe but also may further increase the effectiveness of pharmacological treatments for epilepsy and likely reduce mortality risk (Scorza 2015). It is desirable that this message is forwarded to patients by practitioners, official institutions, and patient associations known to be very dynamic in this area.

5.3.2 Vitamin D

Since the work of Dr. R. Kruse in 1968, many authors have reported the presence of bone decalcification (osteomalacia) and fractures accompanying the long-term treatment of epilepsy with various anticonvulsants. Even if it has not been consistently observed, a link between the disease and low blood vitamin D levels was very frequently reported.

As with MS (Section 5.2.2), the prevalence or the severity of epilepsy seem to follow sunshine variations; in general, in relevant subjects such variations are considered to reflect the vitamin D status.

One of the most important observations was made in London, UK, and showed that along 1 year the seizure frequency in adults was lower in summer than in winter (Baxendale 2009). Similarly, in Toronto, Canada, an annual cycle has been observed in very young children, with a greater frequency of epileptic spasms in December and January and a lower frequency in April and May (Cortez et al. 1997). As emphasized by Dr. S. Baxendale, these findings suggest the development of epilepsy treatment using "light boxes," as already used for other disorders of nervous origin (Section 6.1.1.2).

As for patients with MS (Section 5.2.2), the epilepsy prevalence in a population seems to be based on the month of birth of individuals. Thus, in England and Wales, the systematic perusal of registers of all hospital including epilepsy (30,080 patients born between 1938 and 1988) has shown that a greater proportion of patients were born in December or January, whereas a smaller proportion of patients were born in September (Procopio et al. 1997). Very similar results were obtained in Denmark by the same team using a survey of 50,886 patients (Procopio and Marriott 1998).

Seasonal variations have also been demonstrated in patients with epilepsy in studying the incidence of the brain electrical responses to intermittent light stimuli (photoparoxysmal responses) (Danesi 1988). The records, done in summer, showed much less photoparoxysmal discharges than in winter, with intermediate figures recorded in spring. This scarcity of discharges in summer reflected a weaker neuronal excitability.

Curiously, none of these investigators focusing on exploring the influence of sunshine among patients has evoked a possible relationship with vitamin D status. Perhaps these clinicians had not inspected the extensive literature involving vitamin D (often called solar hormone) in various nervous disorders and more specifically in epilepsy (Leray 2015).

5.3.2.1 *Intervention studies*

Long before these observations, Dr. G. Offermann at the Berlin University had begun in 1979 to verify in patients with epilepsy the possible close association between the vitamin D status, the state of the skeleton, and drug treatments (Offermann et al. 1979). That research recommended to supplement sick children with 37–125 μg (1500–5000 IU) per week of vitamin D, according the time of year. At that time, clinicians postulated that the vitamin D deficiency, commonly seen in young patients, originated from the administered drugs, the ketogenic diet (Section 5.3.1), or from external factors (diet, mobility, obesity).

Since these first investigations carried out in different countries, numerous authors have confirmed the effect of anticonvulsants on the vitamin D levels, with some being more active (oxcarbezine) than others (valproate) (Verrotti et al. 2010). The differences are probably related to the more or less potent effect of these molecules on the enzymes of the vitamin D catabolism (Hollo et al. 2014).

It was also noted that a combination therapy or a ketogenic diet (Section 5.3.1) is more detrimental to skeletal health than a monotherapy. Many clinical studies have shown that a supply of at least 400 IU/day of vitamin D is necessary to restore an adequate vitamin status in children (Harijan et al. 2013). Sometimes, a higher supply (1200 IU/day and above) was required (Snoeijen-Schouwenaars et al. 2015). Control of both vitamin D and calcium levels in blood is therefore necessary to verify the effectiveness of that supplementation and to counter the adverse effects of antiepileptic drugs on skeletal formation.

The vitamin D deficiency induced by the epilepsy therapies does not seem to be regularly taken into account by clinicians. That situation is probably due to a lack of clear instructions from medical authorities or pharmaceutical firms, with the latter being reluctant to promote inexpensive products generating only few profits. Notably, the French Neurology Society states on its website that many authors propose at least to ensure an adequate supply of calcium and vitamin D. That academic society suggests also to prescribe a vitamin D supplementation for all patients treated with phenobarbital, phenytoin, carbamazepine, and valproate. In contrast, in the United States, the Pediatrics Academy recommends that all children receive 400 IU/day of vitamin D, but does not specify a precise treatment for children treated with anticonvulsants.

What about the possible effect of vitamin D on the disease itself? In the early 1970s, the observation of a hypocalcemia in patients treated for epilepsy had alerted some researchers, and one of them had the idea to check whether a vitamin D supplementation could correct this state and also influence the number of seizures. Thus, for the first time in 1974, a clinician, Dr. C. Christiansen in Glostrup, Denmark, supplemented 23 young epileptic patients aged from 6 to 27 years for 28 days with 4000 IU/day (Christiansen et al. 1974). The study showed that all the patients had originally typical "grand mal" or focal motor attacks and several of them had also other types of seizure. In addition, all patients were also treated with anticonvulsant drugs. It was noted that, despite a persistent hypocalcemia, the seizure frequency was reduced by 30% compared to a group of patients receiving a placebo. It is now more than 40 years later since the author advised that "these results would lend further support to the concept that prophylactic vitamin D treatment is advisable for epileptics on anticonvulsant therapy." Why did this work remain in the shadows for nearly 40 years? No one knows. Are neurologists more reluctant than other specialists to change their concept? Yet, a Hungarian team of the Neurosciences National Institute in Budapest took over and confirmed the results published by Dr. C. Christiansen (Hollo et al. 2012). These researchers administered adult subjects, epileptic since their youth, an initial dose of 40,000–200,000 IU of vitamin D to correct their initial hypovitaminosis, and then 2000–2600 IU/day for 3 months. At the end of the experiment, all subjects had a satisfactory serum vitamin D level (23–45 ng/mL), and the number of seizures decreased, on average, by 40%. Five patients even had a decrease of more than 50%. In one of these patients who had a blood vitamin D level of 4 ng/mL before supplementation and 43.1 ng/mL after, the number of seizures during the 90-day period before and after treatment onset decreased from 450 to 30, respectively.

As the authors pointed out for epilepsy, it is surprising that the consequences of a vitamin D deficiency for epilepsy are still unknown. Is it the result of the conventional treatments, or would it contribute to the disease development? An early response on the possible effects of vitamin D seems to be provided by electrophysiological experiments in animals where, as in humans, the same effects on a drug-induced epilepsy were observed. Indeed, this vitamin behaves in mice as a true antiepileptic drug; furthermore, it is able to enhance the action of therapeutic substances, such as valproate and phenytoin, without changing their concentration in the brain (Borowicz et al. 2007). In Greece, a confirmation of that mechanism seems also to be provided by an investigation done in newborns. The latter, only breast-fed, showed clearly that in the presence of epileptic symptoms, doctors must always look for a vitamin D deficiency accompanying an early rickets, especially if the mother is also deficient (Mantadakis et al. 2012). This association between vitamin D and epilepsy has been also verified in Turkey with children aged from 5 to 16 years (Sonmez et al. 2015). Comparing throughout the year control children with 60 children recently diagnosed with idiopathic epilepsy but receiving no drug, it has been found that according to season they had serum vitamin D levels 30%–39% lower than control children. These studies clearly show a regular combination of vitamin D deficiency with epilepsy. It seems therefore that a good practice remains to counteract any deficiency with a well-controlled vitamin D supplementation.

It is hoped that further research will clarify these aspects of prevention and treatment of this neurological disease with a vitamin initially known to help the building of the skeleton. It is obvious that more work is necessary to improve the knowledge of this pathology, affecting at least 3 million patients in Europe. It is unfortunate that in 2016, only two clinical trials on the relationships between vitamin D and epilepsy were declared worldwide (www.ClinicalTrials.com).

In conclusion, it seems that right now the preventive and therapeutic effects of vitamin D on epilepsy must be taken seriously, especially when one considers all the work done in animals and all the clinical, "ecological," or experimental studies. As emphasized by Dr. A. Hollo of the National Institute for Medical Rehabilitation in Budapest, Hungary, low levels of circulating vitamin D caused by the most conventional treatments of epilepsy and the widespread deficiency in that vitamin advocate in favor of a systematic supplementation in patients with epilepsy.

References

Agim, Z.S., Cannon, J.R. 2015. Dietary factors in the etiology of Parkinson's disease. *Biomed. Res. Int.* 2015:672838.

Akhtar, S., Alroughani, R., Al-Shammari, A., et al. 2015. Month of birth and risk of multiple sclerosis in Kuwait: A population-based registry study. *Mult. Scler.* 21:147–54.

Ascherio, A., Munger, K.L., White, R., et al. 2014. Vitamin D as an early predictor of multiple sclerosis activity and progression. *JAMA Neurol.* 71:306–14.

Baxendale, S. 2009. Seeing the light? Seizures and sunlight. *Epilepsy Res.* 84:72–6.

Borowicz, K.K., Morawska, M., FurmanekKarwowska, K., et al. 2007. Cholecalciferol enhances the anticonvulsant effect of conventional antiepileptic drugs in the mouse model of maximal electroshock. *Eur. J. Pharmacol.* 573:111–15.

Brola, W., Sobolewski, P., Szczuchniak, W., et al. 2016. Association of seasonal serum 25-hydroxyvitamin D levels with disability and relapses in relapsing-remitting multiple sclerosis. *Eur. J. Clin. Nutr.* 70:995–9.

Butler, M.W., Burt, A., Edwards, T.L., et al. 2011. Vitamin D receptor gene as a candidate gene for Parkinson disease. *Ann. Hum. Genet.* 75:201–10.

Campenhausen, S.V., Bornschein, B., Wick, R., et al. 2005. Prevalence and incidence of Parkinson's disease in Europe. *Eur. Neuropsychopharmacol.* 15:473–490.

Cardoso, H.D., dos Santos, E.F., de Santana, D.F., et al. 2014. Omega-3 deficiency and neurodegeneration in the substantia nigra: Involvement of increased nitric oxide production and reduced BDNF expression. *Biochim. Biophys. Acta* 1840:1902–12.

Christiansen, C., Rodbro, P., Sjö, O. 1974. "Anticonvulsant action" of vitamin D in epileptic patients? A controlled pilot study. *Br. Med. J.* 2:258–9.

Cortez, M.A., Burnham, W.M., Hwang, P.A. 1997. Infantile spasms: Seasonal onset differences and zeitgebers. *Pediatr. Neurol.* 16:220–4.

Dahlin, M., Hjelte, L., Nilsson, S., et al. 2007. Plasma phospholipid fatty acids are influenced by a ketogenic diet enriched with n-3 fatty acids in children with epilepsy. *Epilepsy Res.* 73:199–207.

Danesi, M.A. 1988. Seasonal variations in the incidence of photoparoxysmal response to stimulation among photosensitive epileptic patients: Evidence from repeated EEG recordings. *J. Neurol. Neurosurg. Psychiatry* 51:875–7.

DeGiorgio, C.M., Miller, P.R., Harper, R., et al. 2015. Fish oil (n-3 fatty acids) in drug resistant epilepsy: A randomised placebo-controlled crossover study. *J. Neurol. Neurosurg. Psychiatry* 86:65–70.

de Rijk, M.C., Breteler, M.M., den Breeijen, J.H., et al. 1997. Dietary antioxidants and Parkinson disease. The Rotterdam Study. *Arch. Neurol.* 54:762–5.

D'hooghe, M.B., Haentjens, P., Nagels, G., et al. 2012. Alcohol, coffee, fish, smoking and disease progression in multiple sclerosis. *Eur. J. Neurol.* 19:616–24.

Dobson, R., Giovannoni, G., Ramagopalan, S. 2013. The month of birth effect in multiple sclerosis: Systematic review, meta-analysis and effect of latitude. *J. Neurol. Neurosurg. Psychiatry* 84:427–32.

Duan, S., Lv, Z., Fan, X., et al. 2014. Vitamin D status and the risk of multiple sclerosis: a systematic review and meta-analysis. *Neurosci. Lett.* 570:108–13.

Etminam, M., Gill, S.S., Samii, A. 2005. Intake of vitamin E, vitamin C, and carotenoids and the risk of Parkinson's disease: A meta-analysis. *Lancet Neurol.* 4:362–5.

Evatt, M.L., Delong, M.R., Khazai, N., et al. 2008. Prevalence of vitamin D insufficiency in patients with Parkinson disease and Alzheimer disease. *Arch. Neurol.* 65:1348–52.

Eyles, D.W., Burne, T.H., McGrath, J.J., et al. 2013. Vitamin D, effects on brain development, adult brain function and the links between low levels of vitamin D and neuropsychiatric disease. *Front. Neuroendocrinol.* 34:47–64.

Forsgren, L., Beghi, E., Oun, A., et al. 2005. The epidemiology of epilepsy in Europe—a systematic review. *Eur. J. Neurol.* 12:245–53.

Goldberg, P. 1974. Multiple sclerosis: Vitamin D and calcium as environmental determinants of prevalence. Part 1: Sunlight, dietary factors and epidemiology. *Int. J. Environ. Stud.* 6:19–27.

Goldberg, P., Fleming, M.C., Picard, E.H., et al. 1986. Multiple sclerosis: Decreased relapse rate through dietary supplementation with calcium, magnesium and vitamin D. *Med. Hypotheses* 21:193–200.

Harijan, P., Khan, A., Hussain, N. 2013. Vitamin D deficiency in children with epilepsy: Do we need to detect and treat it? *J. Pediatr. Neurosci.* 8:5–10.

Hollo, A., Clemens, Z., Kamondi, A., et al. 2012. Correction of vitamin D deficiency improves seizure control in epilepsy: A pilot study. *Epilepsy Behav.* 24:131–3.

Hollo, A., Clemens, Z., Lakatos, P. 2014. Epilepsy and vitamin D. *Int. J. Neurosci.* 124:387–93.

Jelinek, G.A., Hadgkiss, E.J., Weiland, T.J., et al. 2013. Association of fish consumption and Ω 3 supplementation with quality of life, disability and disease activity in an international cohort of people with multiple sclerosis. *Int. J. Neurosci.* 123:792–801.

Kidd, P.M. 2000. Parkinson's disease as multifactorial oxidative neurodegeneration: Implications for integrative management. *Altern. Med. Rev.* 5:502–29.

Kishimoto, Y., Radin, N.S., Tourtellotte, W.W., et al. 1967. Gangliosides and glycerophospholipids in multiple sclerosis white matter. *Arch. Neurol.* 16:41–54.

Knekt, P., Kilkkinen, A., Rissanen, H., et al. 2010. Serum vitamin D and the risk of Parkinson disease. *Arch. Neurol.* 67:808–11.

Kurtzke, J.F., Beebe, G.W., Norman, J.E. 1979. Epidemiology of multiple sclerosis in US veterans: 1. Race, sex, and geographic distribution. *Neurology* 29: 1228–35.

Larsen, J.P., Riise, T., Nyland, H., et al. 1985. Clustering of multiple sclerosis in the county of Hordaland, Western Norway. *Acta Neurol. Scand.* 71:390–5.

Leray, C. 2015. *Lipids. Nutrition and health.* Boca Raton, FL: CRC Press.

Lux, W.E., Kurtzke, J.F. 1987. Is Parkinson's disease acquired? Evidence from a geographic comparison with multiple sclerosis. *Neurology* 37:467–71.

Mantadakis, E., Deftereos, S., Tsouvala, E., et al. 2012. Seizures as initial manifestation of vitamin D-deficiency rickets in a 5-month-old exclusively breastfed infant. *Pediatr. Neonatol.* 53:384–6.

Mokry, L.E., Ross, S., Ahmad, O.S., et al. 2015. Vitamin D and risk of multiple sclerosis: A Mendelian randomization study. *PLoS Med.* 12:e1001866.

Munger, K.L., Levin, L.I., Hollis, B.W., et al. 2006. Serum 25-hydroxyvitamin D levels and risk of multiple sclerosis. *JAMA* 296:2832–8.

Nordvik, I., Myhr, K.M., Nyland, H., et al. 2000. Effect of dietary advice and n-3 supplementation in newly diagnosed MS patients. *Acta Neurol. Scand.* 102: 143–9.

Offermann, G., Pinto, V., Kruse, R. 1979. Antiepileptic drugs and vitamin D supplementation. *Epilepsia* 20:3–15.

Pantzaris, M.C., Loukaides, G.N., Ntzani, E.E., et al. 2013. A novel oral nutraceutical formula of omega-3 and omega-6 fatty acids with vitamins (PLP10) in relapsing remitting multiple sclerosis: A randomised, double-blind, placebo-controlled proof-of-concept clinical trial. *Br. Med. J.* 3:e002170.

Peterson, A.L., Murchison, C., Zabetian, C., et al. 2013. Memory, mood, and vitamin D in persons with Parkinson's disease. *J. Parkinsons Dis.* 3:547–55.

Pringsheim, T., Jette, N., Frolkis, A., et al. 2014. The prevalence of Parkinson's disease: A systematic review and meta-analysis. *Mov. Disord.* 29:1583–90.

Procopio, M., Marriott, P.K. 1998. Seasonality of birth in epilepsy: A Danish study. *Acta Neurol. Scand.* 98:297–301.

Procopio, M., Marriott, P.K., Williams, P. 1997. Season of birth: Aetiological implications for epilepsy. *Seizure* 6:99–105.

Reda, D.M., Abd-El-Fatah, N.K., Omar, T.S., et al. 2015. Fish oil intake and seizure control in children with medically resistant epilepsy. *N. Am. J. Med. Sci.* 7: 317–21.

Schlanger, S., Shinitzky, M., Yam, D. 2002. Diet enriched with omega-3 fatty acids alleviates convulsion symptoms in epilepsy patients. *Epilepsia* 43:103–4.

Schoindre, Y., Terrier, B., Kahn, J.E., et al. 2012. Vitamin D and autoimmunity. First part: Fundamental aspects. *Rev. Med. Int.* 33:80–6.

Scorza, F.A. 2015. Sudden unexpected death in children with epilepsy: Focus on dietary supplementation with omega-3 polyunsaturated fatty acids. *Epilepsy Behav.* 44:169–70.

Scorza, F.A., Cysneiros, R.M., Arida, R.M., et al. 2008. The other side of the coln: Beneficiary effect of omega-3 fatty acids in sudden unexpected death in epilepsy. *Epilepsy Behav.* 13:279–283.

Shoulson, I. 1993. Effects of tocopherol and deprenyl on the progression of disability in early Parkinson's disease. *N. Engl. J. Med.* 328:176–83.

Sies, H., Stahl, W., Sundquist, A.R. 1992. Antioxidant functions of vitamins. Vitamins E and C, beta-carotene, and other carotenoids. *Ann. N. Y. Acad. Sci.* 669:7–20.

Simpson, S., Blizzard, L., Otahal, P., et al. 2011. Latitude is significantly associated with the prevalence of multiple sclerosis: A meta-analysis. *J. Neurol. Neurosurg. Psychiatry* 82:1132–41.

Smith, M.P., Fletcher-Turner, A., Yurek, D.M., et al. 2006. Calcitriol protection against dopamine loss induced by intracerebroventricular administration of 6-hydroxydopamine. *Neurochem. Res.* 31:533–9.

Snoeijen-Schouwenaars, F.M., van Deursen, K.C., Tan, I.Y., et al. 2015. Vitamin D supplementation in children with epilepsy and intellectual disability. *Pediatr. Neurol.* 52:160–4.

Sonmez, F.M., Donmez, A., Namuslu, M., et al. 2015. Vitamin D deficiency in children with newly diagnosed idiopathic epilepsy. *J. Child. Neurol.* 30:1428–32.

Stollberger, C., Finsterer, J. 2004. Cardiorespiratory findings in sudden unexplained/unexpected death in epilepsy (SUDEP). *Epilepsy Res.* 59:51–60.

Suzuki, M., Yoshioka, M., Hashimoto, M., et al. 2013. Randomized, double-blind, placebocontrolled trial of vitamin D supplementation in Parkinson disease. *Am. J. Clin. Nutr.* 97:1004–13.

Taha, A.Y., Ryan, M.A., Cunnane, S.C. 2005. Despite transient ketosis, the classic high-fat ketogenic diet induces marked changes in fatty acid metabolism in rats. *Metabolism* 54:1127–32.

Taha, A.Y., Burnham, W.M., Auvin, S. 2010. Polyunsaturated fatty acids and epilepsy. *Epilepsia* 51:1348–58.

Thouvenot, E., Orsini, M., Daures, J.P. et al. 2015. Vitamin D is associated with degree of disability in patients with fully ambulatory relapsing-remitting multiple sclerosis. *Eur. J. Neurol.* 22:564–9.

Torkildsen, O., Aarseth, J., Benjaminsen, E., et al. 2014. Month of birth and risk of multiple sclerosis: Confounding and adjustments. *Ann. Clin. Transl. Neurol.* 1:141–4.

Verrotti, A., Coppola, G., Parisi, P., et al. 2010. Bone and calcium metabolism and antiepileptic drugs. *Clin. Neurol. Neurosurg.* 112:1–10.

Vukusic, S., Van Bockstael, V., Gosselin, S., et al. 2007. Regional variations in the prevalence of multiple sclerosis in French farmers. *J. Neurol. Neurosurg. Psychiatry* 78:707–9.

Wang, J., Yang, D., Yu, Y., et al. 2016. Vitamin D and sunlight exposure in newly-diagnosed Parkinson's disease. *Nutrients* 8:142.

Weinstock-Guttman, B., Baier, M., Park, Y., et al. 2005. Low fat dietary intervention with omega-3 fatty acid supplementation in multiple sclerosis patients. *Prostaglandins Leukot. Essent. Fatty Acids* 73:397–404.

Willis, W.A., Evanoff, B.A., Lian, M., et al. 2010. Geographic and ethnic variation in Parkinson disease: A population-based study of US Medicare beneficiaries. *Neuroepidemiology* 34:143–51.

Xiao, Y.F., Kang, J.X., Morgan, J.P., et al. 1995. Blocking effects of polyunsaturated fatty acids on Na+ channels of neonatal rat ventricular myocytes. *Proc. Natl. Acad. Sci. U. S. A.* 92:11000–4.

Zhang, S.M., Willett, W.C., Hernán, M.A., et al. 2000. Dietary fat in relation to risk of multiple sclerosis among two large cohorts of women. *Am. J. Epidemiol.* 152:1056–64.

Zhao, Y., Sun, Y., Ji, H.F., et al. 2013. Vitamin D levels in Alzheimer's and Parkinson's diseases: A meta-analysis. *Nutrition* 29:828–32.

Zimmer, L., Vancassel, S., Cantagrel, S., et al. 2002. The dopamine mesocorticolimbic pathway is affected by deficiency in n-3 polyunsaturated fatty acids. *Am. J. Clin. Nutr.* 75:662–7.

chapter six

Mental disorders

Mental disorders are also known as psychiatric disorders. There are different types that induce, depending on the disease, discomfort in daily life and more or less severe suffering or behavioral disorders. The WHO defines these disorders as "psychiatric illness" or diseases that appear primarily as abnormalities of thought, feelings, or behavior, causing distress or dysfunction. Their manifestations most often appear in adolescence and early adulthood. Among these diverse pathological situations are depression, schizophrenia, bipolar disorder, borderline personality disorder, attention-deficit hyperactivity disorder, and autism spectrum disorder. The majority of clinical studies use a classification system regularly updated, recorded in the Diagnostic and Statistical Manual of Mental Disorders (DSM-IV 1994 revised in 2000), internationally accepted and published by the American Psychiatry Association. In Europe, a close system, the International Classification of Diseases, established by WHO, is also used.

The DSM-IV classes all psychiatric disorders on five axes; only some disorders classified in the first axis are considered here (the major clinical disorders) because their relationships with nutrition becoming actually better known.

According to WHO, mental disorders or psychiatric illnesses in all their diversity are worldwide the largest cause of disability. The cumulative annual cost of mental disorders in the United States represents about 2.5% of the gross national product.

Throughout Europe, and taking account only hospitalizations, the WHO estimated in 2011 that the share in expenditures spent on mental disorders is 20%–25% of the total cost of health services. An evaluation of the size of the concerned European population and the financial burden was performed by Wittchen et al. (2011). Each year more than 38% of the European population suffer from a mental disorder, or approximately 165 million people. Nearly one third of patients currently are receiving a treatment; and given the extent of future needs, the authors have stated that these mental disorders could become a major health problem for the twenty-first century.

In France, the total costs of mental disorders were assessed at about 109 billion euros per year, with the direct health expenditures amounting to 20% and the rest being attributed to social costs (Chevreul et al. 2013).

6.1 Major clinical disorders

Major clinical disorders include depression, bipolar disorder, schizophrenia, attention disorders, and those with autism spectrum. Depression and bipolar disorders are often grouped together under "mood disorders." This list is completed by two types of behavior disorders leading to aggressions or suicide (Section 6.2).

6.1.1 Depressive disorders

Depressive disorders are the most important among the so-called mood disorders. An estimated 121 million people worldwide currently suffer from some form of depression. Depression reportedly affects 1 in 10 Americans, but the incidence is actually higher in some states than in others. Certain ethnicities also report higher depression rates than do others. In 2014, an estimated 15.7 million adults aged 18 years or older had at least one major depressive episode in the past year. This number represented 6.7% of all US adults. Using national survey and administrative claims data, it has been estimated that the incremental economic burden of individuals with major depressive disorder between 2005 and 2010 increased by 21.5% (from US$173.2 billion to US$210.5 billion). That rising societal burden is primarily due to a combination of the population growth, an increasing disease prevalence rate, and higher treatment costs per patient (National Institute for Mental Health). A possible complication of depressive illness in combination with other risk factors is suicide. It has been estimated that up to 15% of those who are clinically depressed in the United States die by suicide. It was also found that suicide rate was highest when the economy was weak. Dr. Alex Crosby, an epidemiologist at the Centers for Disease Control and Prevention, determined that the highest rates were during the Great Depression in 1932: 22.1 per 100,000, about 70% higher than in 2014.

European epidemiological studies revealed that nearly 10% of the population suffer from depression, and one quarter of the population may have a mental disorder during life. In 2011, it was estimated that nearly 30 million people in Europe suffered from these troubles; there were 18.4 million in 2005 (Wittchen et al. 2011). A report of the European Brain Council evaluated the direct costs to 23% of a total of 113 billion euros, with the rest of expenses mainly covering sick days and sometimes the consequences of suicide attempts. For society, the burden is growing steadily as a result of an aging population becoming more vulnerable and with multiple diseases. Forecasters indicate that around 2030, depression could be, in importance, the first disease and the heaviest burden to be supported by all states. Indeed, world experts say that the cost associated with these disorders could quickly bankrupt the national health systems, knowing that mental health problems are already more severe than those related to obesity or cardiovascular diseases.

In France, depression is the most common psychiatric illness; nearly 3 million people are affected (INPES 2007). It is the first cause leading to a support by psychiatric hospitals (20% of adult internees). That pathology is predominant between 25 and 45 years and from adolescence it is two times more common in women than in men. Although the average age of the first depressive episode is between 30 and 35 years, the onset of the disease is possible at any age.

The symptoms of depression (also called unipolar disorders) are highly varied; they are characterized most often by feelings of worthlessness, helplessness, and hopelessness. The diagnosis is established when an episode lasts at least 15 days. The patient is sad; has no interest in life, no appetite, no energy, no sleep; and constantly feels fatigue. These disorders can lead to suicidal thoughts when the subject is in deep distress. It has been estimated that in France depression is the leading cause of suicide: almost 70% of people who die by suicide suffer from depression, often undiagnosed or untreated (http://www.france-depression.org).

Recall that research has demonstrated that pregnant women with depression can pass the condition on to their unborn babies. Thus, MRI scans have suggested that abnormal amygdala function can be transmitted from mothers to babies before birth. The good news is that this risk might be reduced by systematic screening of pregnant women for depression and then initiating effective treatment. This situation emphasized the importance of therapies devoid of chemical compounds, but rather based on dietary control.

Epidemiologists and clinicians measure the extent and intensity of depressive symptoms by using a questionnaire completed by a patient and then evaluated by a doctor. The scales used are all derived from the Center for Epidemiologic Studies Depression (CES-D) scale established in 1977 in the United States by the Center for Epidemiological Studies of the National Institute of Mental Health (Radloff 1977) (Section 7.11).

The goal of current treatments for depression is mainly to reduce symptoms and their impact on daily life, and especially to prevent relapses. Although there are effective treatments, it has been estimated that even under optimal conditions they cannot relieve more than about one third of patients with major depressive disorders. So, regardless of the method used in people at risk, the prevention is only effective in 1 in 5 (Opie et al. 2015).

In general, antidepressant drugs are targeting serotoninergic, norepinephrine, and dopaminergic neurons. In case of failure of two or three successive treatment periods, the chances of success with a new treatment drop to 13%.

In some cases, the depressive period may extend over several years; that is, a chronic depression; but if the symptoms are fewer and less intense, it is a dysthymia.

Given the significantly inadequate results in the fight against depression, it became necessary to consider a new approach, as well for the prevention as for the treatment of current depressive disorders. Many studies on laboratory animals and some clinical trials have shown the way by focusing on the benefits of a Mediterranean diet but also of more specific factors such as ω-3 fatty acids (Section 6.1.1.1), vitamin D (Section 6.1.1.2), and vitamin E (Section 6.1.1.3).

6.1.1.1 *ω-3 fatty acids*

Besides the field of genetics and brain chemical messengers (serotonin, dopamine), how did researchers suspect a possible influence of certain essential fatty acids?

6.1.1.1.1 Epidemiological investigations Smith (1991) and Hibbeln and Salem (1995) already hypothesized that throughout the twentieth century the increased incidence of depressive disorders and their early onset could well be due to the increasing consumption of ω-6 fatty acids by the population. Numerous studies have indeed shown that our diet has been gradually enriched with these compounds, as a result of increasing consumption of vegetable oils rich in ω-6, but very poor in ω-3 (groundnut, sunflower) fatty acids. These major changes in the appearance of new cases of psychiatric diseases are to be compared with the increased incidence of cardiovascular and inflammatory diseases. It is now well known that these latter troubles are aggravated by a ω-3 fatty acid deficiency and especially by an imbalance of the ω-6/ω-3 fatty acid ratio. R. S. Smith thus placed in opposition the proportion of depression in the United States and that described in Japan to be 10 times lower. This comparison suggested a parallel with a higher consumption of fish rich in ω-3 fatty acids in Japan. Using statistical data compiled within nine countries with very different cultures, J. R. Hibbeln determined that the number of cases of major depression was inversely related to the average number of fish consumed by the populations (Hibbeln 1998) (Figure 6.1).

Conversely, based on official statistics from five countries (Argentina, Australia, Canada, US, and UK), Hibbeln et al. (2004a) verified that the amount of ingested linoleic acid, the dominant ω-6 fatty acid in the Western diet, was positively related to homicide rates. It would be interesting to generalize these relationships to other people to follow their evolution in some countries where health statistics are reliable.

This hypothesis seems consistent with the discovery that a vegetarian diet is frequently associated with a high risk of depressive disorders, without the certitude that these results could be attributed to an excess of ω-6

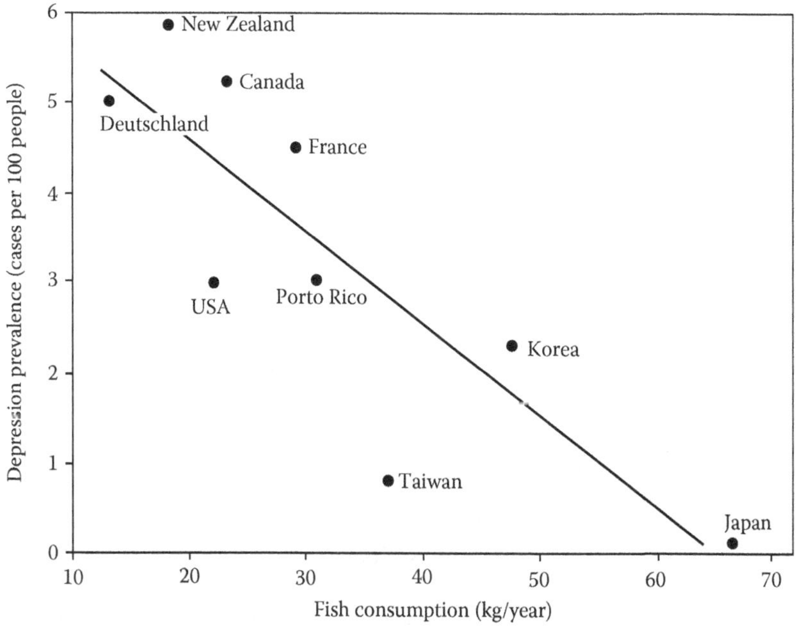

Figure 6.1 Prevalence of depression in nine countries based on their fish consumption. (From Hibbeln, J.R., *Lancet*, 351, 1213, 1998. With permission.)

fatty acids in consumed vegetables. Indeed, it has been shown that the adoption of such a diet was sometimes occurring after the depressive disorder onset (Michalak et al. 2012). It was therefore proposed that their appearance could encourage patients to adopt a vegetarian diet, but other psychological factors could also induce both the occurrence of troubles and a dietary change. With that example, it can be seen that the relationships between diet and mental behavior are very complex. In the future, lengthy investigations with numerous patients would be required to identify the various factors that could motivate the subjects to organize their diet before and during the disease development.

As early as 1983, very high levels of prostanoids deriving from ω-6 fatty acids (prostaglandin E_2 [PGE_2] thromboxane 2) were observed in patients with unipolar or bipolar disorders (Lieb et al. 1983). The authors insisted on a possible causal relationship between PGE_2 and depression, and they also emphasized that prostaglandins of the 2 series could play a role in mood regulation. They suggested that these lipid mediators, as well as catecholamines, could be considered by the theories of the depression. The question was asked by the English team of D. F. Horrobin, exploring the effect of a tricyclic antidepressant (clomipramine) on the vascular action of PGE_2 (Mtabaji et al. 1977). In the absence of older documents, this short story shows, despite everything, that it takes a long time in medicine to develop and defend a hypothesis that is yet supported by sound scientific data.

In 1996 in the laboratory of Prof. A. J. Sinclair in Victoria, Australia, there was perhaps the first convincing evidence of the involvement of ω-3 fatty acids in depression symptoms. Using biochemical analyses and specific neuropsychological tests recognized by the scientific community, this laboratory team determined that the arachidonic acid/eicosapentaenoic acid (EPA) ratio measured in the plasma and red blood cells was positively correlated to the intensity of the clinical symptoms of depression (Adams et al. 1996). Although a correlation does not necessarily indicate causality, these results surprised the experts and encouraged them to develop this new field of clinical research.

Thus, these innovative results prompted the US Department of Health to publish in 2005 a large report of all work on the effects of ω-3 fatty acids on mental health (Evidence Reports/Technology Assessments, No. 116, Agency for Healthcare Research and Quality (US) July 2005, http://www.ncbi.nlm.nih.gov/books/NBK37689/). After analyzing 79 studies, the report concluded that, despite some positive results, "much more research, implementing design and methods improvements, is needed before we can begin to ascertain the possible utility of (foods or supplements containing) ω-3 fatty acids as primary prevention for psychiatric disorders or conditions." The report is clearly aware of the main problem of new investigations in that field and states that "if future research is going to produce data that are unequivocally applicable to North Americans, it will likely need to enroll either North American populations or populations exhibiting a high ω-6/ω-3 fatty acid intake ratio similar to what has been observed in the diet of North Americans." Since then, researchers have fully grasped the problem and delivered a considerable body of evidence that allows for a better understanding of the known effects of ω-3 fatty acids and hope, in a few years, of great progress for the clinical improvement of depression disorders. Despite some caution, seven specialists in that field have published several practical recommendations for the prevention of depression (Opie et al. 2015). These researchers have stressed the advantage of some traditional diets, such as those characterizing the Mediterranean countries, Norway, and Japan, with all of these diets ensuring a high ω-3 fatty acid supply.

The relationship between consumption of marine fish rich in ω-3 fatty acids and the incidence of depressive symptoms have been and still is the foundation of many investigations. Thus, the analysis of fatty acids in blood and adipose tissue enabled to link high EPA and docosahexaenoic acid (DHA) concentrations with a low incidence of depression symptoms. The study of a population in Bordeaux, France, with mean individual age of 74.6 years, has clearly revealed that among plasma fatty acids, only low

EPA levels characterized the patients with depression symptoms (Féart et al. 2008). Similar research based on analysis of red blood cells reached comparable conclusions. Despite this consensus about ω-3 fatty acids, and particularly EPA, the cause of that relationship remains unclear: Is it of internal origin? Could it be due to a change in the cellular metabolism? Or is it of dietary origin? In contrast, Dr. S. Tsuchimine has detected that the blood levels of ω-3 fatty acids are not likely to be associated with personality traits in healthy young Japanese subjects (11–19 years) (Tsuchimine et al. 2016).

Among the most compelling work that merits mention is a study on Korean subjects showing that the severity of depression symptoms was inversely related to the ω-3 index (Section 7.1) measured in red blood cells, as well as to markers of inflammation and oxidative stress (Baek and Park 2013). This study focused on the complex links often mentioned between inflammation and depression, with the ω-3 fatty acids having a great influence in the fight against these two important aggressive phenomena. Even in elderly subjects consuming regularly fish and having high serum ω-3 fatty acid levels, Japanese authors have revealed that depressive symptoms were significantly associated with lower serum levels of DHA, arachidonic acid, and folate (vitamin B_9) (Horikawa et al. 2016).

By 2007, fundamental results were reported by the recognized specialist Prof. R. K. McNamara of the University of Cincinnati, Ohio (McNamara et al. 2007a). Indeed, they showed via postmortem analysis of brain samples from adult subjects with severe depression that the area called the orbitofrontal cortex had lower DHA content (−16% in men and −32% in women) than in control subjects. This brain area is known to be involved in the regulation of hedonistic and emotional processes associated with depression disorders. Given the stability of the composition of brain fatty acids, Prof. McNamara felt that the observed DHA deficiency has yet been established during the perinatal period and could not therefore be corrected later by the diet.

Nutritional assumption seems preferred considering the great number of epidemiological studies worldwide that have shown an inverse relationship between fish consumption and the prevalence of depression. In Europe, two large studies conducted in Finland with thousands of people have supported these findings, but curiously with sharper results in women than in men. The study of many American subjects aged 30–65 years, using questionnaires on their diet and their depression level, showed that the intensity of depression symptoms was inversely related to the amount of ingested ω-3 fatty acids (Beydoun et al. 2013). In Japan, a similar investigation on people aged 23–63 years came to parallel conclusions (Yoshikawa et al. 2015). Furthermore, this study revealed a positive relationship between fish consumption and resilience to depression, an important outcome because it represents the ability of the individuals to overcome the stress face to adversity. Using analyses of dietary lipids, others have used the ω-3/ω-6 fatty acid ratio and come to the same conclusions.

The hypothesis of a possible fight against depression through fish consumption seems strengthened by the latest study initiated by the team of Prof. D. Zhang, Qingdao University, China. This large meta-analysis, covering 26 reliable works published between 2001 and 2014 and gathering together 150,000 subjects, showed that eating fish has a good antidepressant effect, but only in Europeans, with other populations (Americas, Asia, Oceania) not being significantly affected (Li et al. 2016). The reduced size of the cohorts studied in each location outside Europe could be at the origin of this discrepancy. Therefore, it can be concluded that a high fish intake seems beneficial for the primary prevention of depression. Authors have thus determined, on average, a 17% reduction in disease risk of people consuming the highest amount compared with those who consuming the least (20% less risk in men and 16% less risk in women).

It is remarkable that when comparing the populations of several countries, it was observed that the prevalence of depression decreases with increasing fish consumption. Thus, the Asian populations (Japan, Korea, and Taiwan) have the lowest rates of major depression and are also those that consume more fish. These differences could explain the speed with which depression seems to be spreading in the West for 50 years. Indeed, the intake of ω-3 fatty acids is estimated to be now two times lower than it was before World War II. Again, the incidence of depression has greatly increased over the same period.

It has been also suggested that the high fish intake among Icelandic people was responsible for their lack of seasonal depression. Similarly, postpartum depression seems to be less common in women frequently eating fish or having breast milk with high DHA content.

Remember that this type of study cannot bring irrefutable and definitive answers, because in addition to the imprecision of consumption surveys, many genetic, economic, or sociocultural factors could interfere with the physiological data. Moreover, on a practical level, it was shown that the culinary preparation of fish also influences the intensity of depression. Thus, the consumption of breaded fish increased the risk for more severe depression symptoms (Hoffmire et al. 2012). It has been also stressed that the influence of the cooking method, such as roasting in an oven or grilling, could be damaging for EPA and DHA contents (Chung et al. 2008).

Given this promising research, the European Union has initiated a major 5-year project (MooDFOOD) in nine countries to explore mainly the relationships between ω-3 fatty acids and depression (www.moodfood-vu.eu).

6.1.1.1.2 Intervention studies Has it been possible to complete these results by experimental studies to verify the direct effect of ω-3 fatty acids on depression? Were they administered to patients with depression to reduce the clinical signs of the disease? Even if the published results remain equivocal, it is essential to report the most important outcomes.

In 1966, shortly after the detection by P. B. Adams of a relationship between the blood EPA content and the severity of the disease, the administration of EPA, DHA, or both was attempted in patients suffering from depression. A. L. Stoll at Harvard Medical School, Boston, Massachusetts, demonstrated for the first time in a double-blind study against placebo that fish oil had a positive effect in significantly reducing the manifestations of depression outbreaks (Stoll et al. 1999). These exciting results have led the author to declare that ω-3 fatty acids could be the "missing link" among the relationships between cardiovascular diseases and depression (Severus et al. 1999).

That work was taken over in 2002 by B. Nemets at the Ben-Curion University of the Negev, Israel, who confirmed that administration of EPA during 4 weeks strongly decreased depression disorders in more than half of patients with a major depression (Nemets et al. 2002). These encouraging results were popularized in a book (Servan-Schreiber 2003), where the author has spread, in France, the idea of a possible benefit of fish oil intake in restoring the mental balance by formulating the hypothesis of a stabilization of cellular membranes in the brain. However, he wrote that "it will take several years before a sufficient number of studies of this type is achieved. Indeed, the ω-3 fatty acids are natural products and thus almost impossible to patent. Therefore, they do not interest the big pharmaceutical companies that fund most scientific studies of depression." These conclusions, a bit pessimistic, obviously could be applied to all areas of medicine where these lipids could be involved. In contrast, this academic research offers a guarantee for their impartiality toward future users.

A careful review of literature on the subject from 2002 to 2011 shows that authors doubt sometimes the usefulness of treatment with specific diets (nutraceuticals), despite a growing research effort in this area. If one takes into account the conclusions of a recent meta-analysis, almost all clinical studies using ω-3 fatty acids to treat depression symptoms concluded that this treatment was beneficial in adult patients properly diagnosed and also in those who had poorly characterized symptoms (Grosso et al. 2014). Sometimes, the results were more decidedly mixed and demonstrated a positive effect but only after 12 weeks and after using the Clinical Global Impression Scale. This test allowed the investigators to assess, on a 7-level scale, a significant improvement of the disease after a daily supplementation with 1.14 g of EPA and 0.6 g of DHA (Park et al. 2015).

Concerns also were issued in 2010 by Dr. K. M. Appleton after analyzing 29 supplementation trials against placebo. However, the author recognized the importance of a ω-3 fatty acid fortification in patients with a severe, but correctly diagnosed depression. Unfortunately, the heterogeneity of the analyzed trials made their comparison mathematically difficult. A more recent meta-analysis of 26 published studies led to the conclusion that ω-3 fatty acid supplementation resulted in a small-to-modest benefit for depressive symptomology (Appleton et al. 2016). Although no differences were found between a treatment with fatty acids versus antidepressants, the authors stated that the "evidence is limited and highly heterogeneous, resulting in findings that are imprecise and potentially biased." Nevertheless, the authors accept that "given the high rates of adverse events associated with some antidepressants, n-3PUFAs may offer an alternative treatment of possible benefit and reduced side effects, but more evidence regarding both the potential positive and negative effects of n-3PUFAs for major depressive disorder is required before such a suggestion can be advocated."

To clarify the responsible natural agent of the observed effects, the quality and the amount of fatty acids used as supplements in several meta-analyses have been rigorously examined. One meta-analysis examined 15 trials published in many countries and highlighted a specific and beneficial effect of EPA on depression. The best positive results were detected only when this fatty acid represented more than 60% of the ω-3 fatty acid mixture, the effects being perceived when the intake was between 0.2 and 2.2 g/day (Sublette et al. 2011). The specificity of EPA in relation to depression was recently verified by Dr. H. Mozaffari-Khosravi, University of Tehran, Iran, by comparing the effects of a daily administration of 1 g of EPA or DHA for 4 months in patients with mild-to-moderate depression (Mozaffari-Khosravi et al. 2013). In 2010, because significant research in this area was noted, the American Psychiatric Association (Steering Committee on Practice Guidelines) recommended EPA-rich supplements as adjunctive therapy for major depressive disorders (http://psychiatryonline.org/guidelines).

Right now, it is well established that the beneficial effects of the consumption of marine fish or fish oil on depression are due to their content in ω-3 fatty acids, EPA in particular, but not to other components such as vitamin D, with the latter being active elsewhere (Section 6.1.1.2). Recall that an essential fatty acid supply may be beneficial not only for mood but also for the cardiovascular system (Sher et al. 2010).

In the case of perinatal depression (or postpartum depression), the majority of the work indicates a more or less pronounced tendency of a direct link between the trouble frequency and the ω-3 fatty acid deficiency. Work based on a survey of 100 Brazilian women highlighted that the prevalence of these troubles was 24% higher when the dietary ω-6/ω-3 fatty acid ratio exceeded a value of 9 (Da Rocha and Kac 2012). These results were so important that a Norwegian team checked their validity soon after the publication by using the fatty acid composition of red blood cells (Markhus et al. 2013).

The most efficient ω-3 fatty acid is still unknown, although EPA and DHA are good candidates; however, it seems wise to consider the recently acquired results by advising fertile or pregnant women to ensure an adequate nutritional intake of these compounds (Section 7.2). If necessary, it is sufficient to increase the consumption of fatty fish or use dietary supplements. Obviously, this advice is aimed primarily at pregnant women having other psychological risk factors.

The relationships between postpartum depression and nutrition is clearly insufficient when considering the prevalence of mental disorders in pregnant women (10%–15%) and their impact on the behavioral, cognitive, and psychomotor development of children (Kingston et al. 2012).

More specific findings on the control of depression disorders could be drawn only with investigations conducted on larger samples and with suitable psychometric or biochemical analyses. The selection criteria should also be subject to a rigorous selection by psychiatry specialists to study the populations that would best benefit from a supplementation with ω-3 fatty acids. This area interests many research teams because 12 large trials reported in the WHO International Register of Clinical Trials are recruiting in several countries (http://apps.who.int/trialsearch/), whereas 96 were registered in 2016 at http://www.clinicaltrials.gov.

The influence of ω-3 fatty acids on depression is part of the great MooDFOOD study led by the University of Amsterdam, The Netherlands, and supported by the European Commission (www.moodfood-vu.eu) in nine countries. This major multidisciplinary project started in 2014 and is expected to last 5 years, with the aims to define relationships between diet, depression, social environment, and obesity.

What is known about the mechanisms involved in the action of ω-3 fatty acids on depression? Much research in animals allowed us to perceive the mechanisms underlying the behavioral troubles induced by inadequate intakes of EPA and DHA. For example, it was found in rats that DHA deficiency induced troubles characterized by a dopamine deficiency in cortical areas, with the reduction likely related to a cognitive deficit. Significant changes were also observed at the level of serotonergic and cholinergic systems. Several neurophysiological studies have suggested that these changes in the synaptic transmission could result from a direct action of fatty acids on the expression of genes involved in neurotransmission processes, membrane plasticity, and neurogenesis. It has been recently discovered that EPA can increase the release of serotonin from presynaptic neurons, whereas DHA influences directly the serotonin receptors of the postsynaptic neurons, with these two mechanisms also being modulated by vitamin D (Section 6.1.1.2) (Patrick and Ames 2015). Certainly, these fundamental findings are at the beginning of a large development in the field of prevention and fight against depression.

Combined with the previous hypothesis, there is also a growing evidence of an action of the inflammation mediators that interfere in maintaining or worsening the depression symptoms (Pascoe et al. 2011). Thus, depressed patients have high levels of specific cellular mediators such as proinflammatory cytokines (interleukins, interferons). These links between depression and cytokines have been the subject of a recent clinical application in patients with chronic hepatitis C in which a supplementation with EPA could prevent depression accompanying the treatment with interferon-α (Su et al. 2014). Research in animals has confirmed these relationships.

The assessment of the participation of specific ω-3 fatty acids in the control of brain oxidative stress may provide additional insight into depression symptoms associated with inflammatory processes. This approach was undertaken in a large cooperative study (Boston Pueto Rican Health Study) that showed an inverse relationship between the ω-3 index measured in erythrocytes and the severity of depressive symptoms, but only in participants with elevated stress biomarkers measured in urine (Bigornia et al. 2016). These observations suggest that the status of oxidative stress may identify those who might benefit from ω-3 fatty acid supplementation to improve depressive symptoms.

Thus, it is certain that ω-3 fatty acids are involved, as well as vitamin D (Section 6.1.1.2) and vitamin E (Section 6.1.1.3), in the regulatory mechanisms generating depression due to their anti-inflammatory function, a function already well known in the cardiovascular domain (Pascoe et al. 2011).

Clearly, larger clinical studies are needed to determine what the best preventive treatments are to curb the symptoms of depression, a pathology generating major social problems in all populations.

Regarding self-medication, all experts agree to warn people with depression or other mood disorders not to use ω-3 fatty acid supplementations without a medical opinion. Such conditions should be analyzed by a competent specialist, because they can be improved but also worsen and become worrisome after some time. If necessary, the first intention will be to find a diet covering properly the ω-3 fatty acid needs, for example, two or three fish meals per week (Section 7.2).

People with mood disorders and treated could, of course, change their treatment, but only after medical information. A ω-3 fatty acid intake cannot replace an antidepressant medication or psychotherapy, but it may improve their efficiency. A doctor's consultation is also essential for pregnant and lactating women and for people with coagulation problems. For now, it seems wise to select a ω-3 fatty acid supplementation only in patients deficient in such compounds, with the deficiency being determined by the study of their usual diet or especially a blood cell analysis. Thus, it seems appropriate to extend the estimation of the ω-3 fatty acid status as it has been proposed with the "ω-3 index" to determine the cardiovascular risk (Von Schacky 2010).

In general, all past basic research has proved that the consumption of foods rich in ω-3 fatty acids is beneficial for the prevention of mood disorders. Unfortunately, these recommendations are insufficiently disseminated in specialized media. For subjects with depressive disorders, a treatment with a mixture of EPA and DHA (1 g/day) is already recommended by many clinicians. Such a treatment can be applied isolated or better, with conventional therapies, thus improving their efficiency. Along with the observance of ω-3 fatty acid intakes, it is recommended to keep sufficient dietary vitamin D and vitamin E intakes. Future research should optimize the type of treatment and the useful dose of these natural products to prevent but also to treat at the onset of the disease.

6.1.1.2 Vitamin D

Besides the effects of vitamin D on cognitive impairment (Section 4.2.3), it seems increasingly evident that a deficiency in this vitamin is involved in the development and the severity of depression disorders.

The correlation between depression and the activity of neurotransmitters (serotonin, norepinephrine, and dopamine) is now well established. Indeed, it is well known that they participate in the regulation of emotional activity, stress reaction, regulation of sleep cycles, appetite, and many others functions. It seems therefore logical that vitamin D, such as antidepressants, acts on the regulation of the balance of the neurotransmitters involved in depression. Recent experimental evidence has confirmed this hypothesis by showing in animals (Jiang et al. 2014) and humans (Kaneko et al. 2015) that vitamin D stimulates the biosynthesis of cerebral serotonin.

This "solar hormone," a real "panacea," might as well be a link in a skin–brain pathway equivalent to the well-known retina-midbrain-epiphysis neuroendocrine pathway, according to a hypothesis reported very early by the famous American specialist of vitamin D target cells Professor W. E. Stumpf (Stumpf 2012).

6.1.1.2.1 Epidemiological investigations It was only in 1984 that light therapy in clinical psychiatry was first used for treating seasonal depression. We owe this discovery, published in 1984, to Dr. E. N. Rosenthal in the United States. The idea that low serum vitamin D levels could be related to depression has derived originally from the observation of a high incidence of seasonal affective disorders (winter depression) when sun exposure is reduced in autumn and winter. An experiment conducted in 1999 by F. M. Gloth in the United States showed that an administration of vitamin D (100,000 international units [IU]) was more efficient than light therapy to treat seasonal depression (Gloth et al. 1999). Nonetheless, the relationships between vitamin D and melatonin are not yet well known, with melatonin being a hormone produced by the pineal gland, mainly in response to the absence of light. More investigations would be required to explore the relationships between light, neurohormones, and vitamin D.

Many researchers were invested in the relationships between vitamin D and depression, encouraged in that direction by the high incidence (up to 45%) of this pathology in people living in nursing homes, thus frequently receiving little sun exposure. To get a clear idea of the magnitude of that disease, check the results of the French PAQUID study, conducted since 1989, on a cohort of about 2800 people over 65 years old living in the Aquitaine region. The prevalence of depressive symptoms in these people, identified from responses to an adapted questionnaire (CES-D scale, Section 7.11), is about 16%, whereas the prevalence among people over 65 years is naturally much lower (2%–3%). Moreover, experts agree that detecting and treating depression is the best prevention of suicide risk in the elderly.

It is possible to find some work on this topic; however, they are unconvincing because they rely generally on a limited number of subjects with uncontrolled lifestyle or diet. Other studies, performed in more or less medicalized nursing homes are more satisfactory because they benefit from better control. The difficulty of comparing results often comes from the type of populations and the diversity of the psychological tests used in different countries to assess the gravity of the current depression. Thus, some epidemiological studies have shown no association between plasma vitamin D and depression symptoms; nothing actually may indicate the reason for these negative results. By contrast, the simple analysis of vitamin D dietary sources together with the clinical observation of depression symptoms in important groups (or cohorts) of the elderly has shown a close relationship between these two parameters. This was the case for a US study (Bertone-Johnson 2009). The outcomes have clearly shown that subjects ingesting more than 800 IU/day (20 µg/day) of vitamin D had 21% lower depression risk compared to subjects ingesting 100 IU/day (2.5 µg/day).

If one considers not only the food intake but also the blood vitamin D level, a parameter recognized as one of the safest to establish a vitamin status, several well-conducted studies have shown a close relationship between depression and vitamin D in adults of any age. An example of this was given in the study published in the United States as part of the Third National Health and Nutrition Examination Survey (Ganji et al. 2010). This survey, conducted on 7970 subjects from 15 to 39 years old, found that those with a serum vitamin D level above 30 µg/L (78 nmol/L) presented significantly less depression episodes than those with a level lower than 20 µg/L (52 nmol/L).

Many other epidemiological surveys carried out in different countries with the elderly achieved similar outcomes; an association between depression and inadequate serum vitamin D levels was always observed.

In early 2015, three published studies still confirmed this trend: one study in Japan with 1786 employees from 19 to 69 years old (Mizoue et al. 2015); one study in Finland with 5371 subjects from 30 to 79 years old (Jääskeläinena et al. 2015), and one study in France with 82 patients with an average age of 46 years (Belzeaux et al. 2015). The Finnish study concluded that maintaining a serum vitamin D level higher than 20 µg/L could reduce the prevalence of depression troubles by 19%. Thus, this effect would have a favorable impact on the related spending caused by depression in Finland (1 billion euros for a country of 5.5 million inhabitants). It must be stressed that the status of vitamin D is still lower than the recommendations of the French Academy of Medicine (Section 7.4).

In France, how much could we save with the application of this accepted biological standard? Probably about 6 billion euros on an expenditure of 30 billion in 1992 caused by only the treatment of depression (CREDES). Why is this financial aspect not taken into account by the relevant associations, the academic institutions or the Ministry of Health?

Besides these studies, a recent study of a large and diverse Danish population, aged 18–64years, did not reveal any link between the vitamin D status, the self-reported symptoms, and diagnosis of depression or anxiety (Husemoen et al. 2016). The small range of vitamin D concentrations (17–28 ng/mL) and their low levels (insufficiency zone) might be at the origin of these conclusions.

In Japan, the results of a study conducted in 2009 should warn researchers about the period of blood collection within the year (Nanri et al. 2009). Indeed, the association between low blood vitamin D levels and deep depression was much more pronounced when the subjects were examined in November compared to subjects having their blood sampled in July. A similar result was obtained in a study of young American students that observed a smaller number of depressive subjects in autumn compared to in winter or spring, with these subjects also having higher vitamin D levels (Kerr et al. 2015).

These last results are consistent with the hypothesis of a protective effect of vitamin D on depression symptoms known to be seasonal troubles and strongly related to the synthesis of vitamin by exposing the skin to sunlight (Gloth et al. 1999).

What are the biological mechanisms by which vitamin D level may predispose to depression? Vitamin D has been shown to modulate serotonin production and glucocorticoid-induced hippocampal cell death. Furthermore, it has been hypothesized that it plays a role in the production of neurotrophins and acetylcholine and in the regulation of specific calcium channels. Vitamin D may also influence depressive symptoms via its proposed anti-inflammatory effect.

Despite these findings, many experiments tend to confirm that depression is not primarily due to an unbalanced diet or to a reduced sun exposure, with both being possible causes of reduced sources of vitamin D. It is obvious that supplementation experiments could only demonstrate a clear causal link between vitamin D and depression after the recognition of a reduction of symptoms, leading then to consider a simple and natural fight against that disease. The links between depression and ω-3 fatty acids (Section 6.1.1.1) or vitamin E (Section 6.1.1.3), now better and better known, should be also considered in research protocols, so as to reduce the dispersion of results by controlling the maximum number of physiological parameters.

Given this promising research, in 2014 the European Union initiated the major 5-year project MooDFOOD in nine countries to explore the relationships between vitamin D and depression, among other factors (www.moodfood-vu.eu).

6.1.1.2.2 Intervention studies In 2004, one of the first studies was published that reported the effects of a 6-month vitamin D supplementation on its blood level but also on mood and well-being, with the latter being estimated with parameters such as energy level, sleep, interest, pleasure, attention, and weight loss (Vieth et al. 2004). The authors showed that beneficial and significant effects were already observed with a dose of 600 IU/day (15 μg/day), but they were even more obvious with a dose of 4000 IU/day (100 μg/day). It also seemed that a regular intake of vitamin D is more efficient than the one-time administration of high doses.

A meta-analysis of seven studies, selected by their strict protocol and published from 1998 to 2013, showed that the restoration of blood vitamin D levels between 17 and 33 μg/L improved depression levels significantly (Spedding 2014). A recent trial conducted at Tehran University, Iran, confirmed that a weekly intake of 50,000 IU (1.25 mg) improved the depression symptoms substantially in deficient adults after 8 weeks (Sepehrmanesh et al. 2016).

These findings show that the effects of treatment of depression by vitamin D are similar, in general, to those obtained with conventional antidepressant drugs. If these results are confirmed in several other large studies, the findings summarized above could have quick implications for public health policy concerning a widespread pathology.

This area of research is currently interesting several research teams worldwide, because 91 clinical trials were registered in 2016 (http://www.clinicaltrials.gov).

All of the most recent work concludes that there is a correlation between vitamin D and depression, however, without being able to define the physiological mechanisms. To support this demonstration, several studies involving supplementation have shown that, in general, symptoms of depression or mood disorders are linked to a vitamin D deficiency, with the latter resulting from a too low dietary intake, an inadequate sunlight exposure, or both. In the current state of knowledge, experts suggest that a supplementation with vitamin D may be beneficial for depressed subjects, especially if they are clearly deficient or if their lifestyle and their place of residence suggest a potential risk of permanent deficiency. No official medical recommendation has been proposed so far for this type of treatment. As for ω-3 fatty acids, the monitoring of blood vitamin D levels is basic. A deficiency should be the subject of a medical consultation, followed possibly by a dietary change, an increasing sun exposure, or both.

6.1.1.3 Vitamin E

In 1962, Dr. F. Post suggested that a cerebrovascular disease was at the basis of affective disorders as observed in about 12% of the elderly (Post 1962). Thus, many investigators have raised the possibility that the generation of free radicals, consecutive to the natural oxidation of membrane lipids, could play a role in the onset of neuropsychiatric disorders such as depression. These free radicals are produced during inflammatory episodes, during immune reactions, or during the catabolism of monoamines (dopamine, adrenaline, and serotonin). This assumption was also raised for ω-3 fatty acids, with the latter being both antidepressant and anti-inflammatory (Lu et al. 2010). Some research seems to confirm this hypothesis. Thus, an increased lipid oxidation together with more intense activities of antioxidant enzymes has been observed in patients with major depressive disorder. Conversely and more surprisingly, an antidepressant treatment (with a serotonin reuptake inhibitor) is able to decrease lipid oxidation (Bilici et al. 2001).

This close relationship between depression and lipid oxidation seems to be confirmed by the demonstration that serum vitamin E levels are significantly lower in patients with major depression compared to healthy subjects (Maes et al. 2000). Two years later, doubts have yet been issued from a Dutch study with a larger number of subjects (262 depressed and 459 controls), taking into account many factors likely to interfere with the interpretation of results (Tiemeir et al. 2002). Indeed, in all subjects, the challenge resided in the monitoring of factors such as depression severity, disability, smoking, diet, and social level, all factors known to alter blood antioxidant levels. The authors emphasized the difficulty of attributing a meaning to the circulating vitamin E levels. Thus, if a lonely person is unable to do his or her shopping, is he or she depressed because of lifestyle or because of an unbalanced diet inducing a vitamin E deficiency?

According to Dr. A. J. Owen, Wollongong University, Australia, the dietary vitamin E intake does not seem to act directly; he observed that the severity of depression disorders was linked to low blood α-tocopherol levels, even with an adequate vitamin supply (Owen et al. 2005). That hypovitaminosis could therefore result from an increased use of vitamin E, along with an increased oxidative stress accompanying depression.

Much more research is needed before understanding the nature of the links between depression and vitamin E.

In conclusion of all these studies and in the absence of well-controlled clinical trials with large cohorts, any important supplementation of vitamin E may not be recommended preventively or curatively in the context of depression disorders. In addition, future trials should take into account the personalized assessment of an oxidative stress, a difficult task when one knows the number of the involved molecules. Trials have been already proposed by some private laboratories. Maintaining a diversified diet consistent with the official recommendations, providing about 15 mg of vitamin E per day up to an old age, is now the only advice that may be delivered to the population. At the same time, it is recommended to maintain adequate dietary ω-3 fatty acids and vitamin D intakes, micronutrients that are likely involved in the depression by increasing the brain serotonin secretion.

6.1.2 *Bipolar disorder and ω-3 fatty acids*

Bipolar disorder (formerly manic depression) is characterized by mood disorders defined by repetitive fluctuations between periods of marked excitation (intense excitement or euphoria that may last for weeks or months) and periods of melancholy (extreme depression with pathological sadness). Between these two episodes, the subject is again in a calm period returning to the usual state. Some severe cases may lead to dangerous behavior, even suicide (8%–10% of deaths by suicide in these patients). Two main types of bipolar disorder have been described; they are distinguished by the presence of a distinct period of at least 1 week of intense or irritable mood (mania, type I) or a period with milder levels of mania, known as hypomania (type II) where individuals are energetic, excitable, and productive. Many patients with bipolar disorder exhibit psychotic symptoms that may be confused with schizophrenia or other mood disorders. During clinical trials, this uncertainty may lead to a dispersion of results and a bias in their interpretation.

Worldwide, bipolar disease affects more than 1% of adults and globally is part of the 10 most costly and debilitating diseases. In the United States, the prevalence of the disease is increasing in adults, rising from 905 patients per 100,000 in 1995 to 1679 per 100,000 in 2003. This increase is even higher among young people under 20 years old.

In 2011, it was estimated that in Europe nearly 3 million people suffered from the disease; there were 2.4 million in 2005 (Wittchen et al. 2011).

In France, more than 600,000 people of all ages and both sexes could be affected; it has been estimated that the prevalence is such that 8% of individuals will present these disorders during their lives. According to psychiatrists, 50% of depressions diagnosed during consultations are actually affected by bipolar disorder. If the intensity of troubles may decrease over time, the disease will persist for long. This pathology has important physical, social, financial, and professional consequences because many bipolar patients are unable to work.

Bipolar disorder has been deemed the most expensive behavioral healthcare diagnosis, costing more than twice as much as depression per affected individual. In the United States, an estimate of the total cost of bipolar disorder published more than a decade ago was as high as US$45 billion/year.

The estimated UK national cost of bipolar disorder was £4.59 billion, with hospitalization during acute episodes representing the largest component of the cost.

In France, studies have estimated that the annual cost of hospitalizations represents 1.3 billion euros.

Although this pathology is partly genetic, it is also recognized that environmental factors may be involved. Among these factors, the quality of dietary lipids has been repeatedly mentioned.

6.1.2.1 Epidemiological investigations

A multicenter comparative study of the bipolar disorder prevalence according to fish consumption was conducted in the United States by the team of Dr. J. R. Hibbeln, a recognized specialist on the relationship between ω-3 fatty acids and psychiatric disorders at the National Institutes of Health (Bethesda, MD) (Noaghiul and Hibbeln 2003). As for depression, this work has clearly shown that the prevalence of the disease decreased according to an exponential decay when fish consumption increased from 5.4 to 100 kg/year. This is the case in Iceland where the prevalence is about 30 times lower than in Germany. It must be emphasized that below a consumption of about 23 kg/year, prevalence was increasing rapidly. That consumption corresponds to about 440 g/week (two large fish servings). The amount of fish consumed would thus correspond to a minimum of about 1.2 g/day of EPA plus DHA calculated from salmon, about twice the amount usually recommended by the medical organizations (0.5 g/day) for the prevention of cardiovascular disease (Leray 2015).

As for other mood disorders, it is significant that the analysis of specimens of prefrontal cortex sampled during autopsy in patients who have suffered from bipolar disorder has shown lower DHA concentrations than in control individuals (McNamara et al. 2007b). This DHA deficiency has been also observed after analyzes of red blood cells (McNamara et al. 2010) or plasma (Pomponi et al. 2013) in patients compared to control subjects. Unfortunately, the question whether this well-defined deficiency could be connected to a reduced food intake of ω-3 fatty acids has not been included in these studies, other studies rather supporting a reduced DHA biosynthesis (Clayton et al. 2008). From these examples, it seems that no one yet knows whether this apparent DHA deficiency is the cause or the consequence of the disease, or whether it could have originated from the pharmacological treatments. Unfortunately, this lack of knowledge can only delay any progress in this domain.

6.1.2.2 *Intervention studies*

What may we learn from the therapeutic attempts of ω-3 fatty acid supplementation? In 1999, Prof. A. L. Stoll, Harvard University, Boston, Massachusetts, conducted one of the first nutritional experiments in patients with bipolar disorder. After 4 months, he observed a significant beneficial effect with a daily prescription of 6.2 g of EPA and 3.4 g of DHA (fish oil), with patients also having psychotropic or psychotherapeutic treatments (Stoll et al. 1999).

After this publication, four in five studies, using a single EPA supplementation, resulted in similar conclusions. It seems that a daily intake of 1 g of EPA is sufficient to notice after only 3 months of treatment beneficial effects in patients taking also conventional drugs (Frangou et al. 2006). Encouraging results have even been achieved in children aged 6–17 years with bipolar disorder and treated with an EPA-rich oil. These results are even more interesting as pediatricians know that the treatment of such young subjects still poses several problems (Wozniak et al. 2007).

The safety of EPA and the absence of any side effect allow to conceive an efficient background treatment of bipolar disorder, accompanying any other therapy chosen by the physician. Despite the lack of precise data, one can hypothesize that a ω-3 fatty acid supplementation as an adjuvant allows a more or less rapid reduction of the psychotropic doses, a condition that should attract the interest of clinicians and patients.

A recent meta-analysis based on five studies comparing a treated group with a placebo group, all with at least 15 subjects, concluded that fish oil consumption has a beneficial effect on bipolar symptoms, with the effect greater for phases of depression than for phases of mania (Sarris et al. 2012). All these studies have shown that ω-3 fatty acid intake (especially DHA and EPA) is the most active, with their metabolic precursor linolenic acid having no effect. The specificity of EPA gives similar results obtained in subjects treated for depression.

The mechanisms involved in the study of ω-3 fatty acid effects on bipolar disorder are probably similar to those explored for depression, with some authors having also discussed a possible action of some lipid mediators, mainly the endogenous cannabinoids (anandamide, 2-arachidonoyl glycerol).

This area of research is currently interesting several research teams worldwide because about 20 clinical trials in children and adolescents were registered in 2016 at http://www.clinicaltrials.gov.

The whole results obtained in the field of psychiatric disorders seemed so convincing that the American Psychiatric Association decided in 2006 to recommend to patients with psychiatric troubles, especially unipolar and bipolar disorders, to consume at least 1 g/day of an EPA plus DHA mixture, or to eat at least three fish servings per week. These natural products display only negligible biological risks and are furthermore beneficial for patients who have also cardiovascular risks. The cost-effectiveness of such treatment is also profitable in the current psychiatric practice.

6.1.3 Schizophrenia (psychotic disorders)

In 1896, the German psychiatrist Emil Kraepelin established a pathological entity he called "early dementia," consisting of a combination of three situations: disorganized behavior, negativistic personality, and paranoid dementia (rich, illogical, and incomprehensible delirium). In 1911, the disease was specified and named "schizophrenia" by the Swiss psychiatrist Eugen Bleuler. This severe psychosis occurs most often in early adulthood, involving delusions, hallucinations, disorganized thinking and speech, behavioral disorders, cognitive disorders, and various chronic symptoms constituting sometimes a serious handicap. That pathology should not be confused with the dissociative identity disorder. All these symptoms usually result in a loss of contact with the world outside, sometimes with an autistic withdrawal. Some patients have simultaneously symptoms of schizophrenia and bipolar disorder, constituting the "schizoaffective" disorder. The closeness of the symptoms may obviously reflect a possible transition between the two diseases, likely explaining the difficulty of the investigations and the uncertainty of some conclusions from often dispersed results.

In general, the symptoms can be grouped into two categories: positive symptoms, described as excess or distortion of normal functions (such as hallucinations and delusions) and negative symptoms characterized by a deficit of normal emotional responses and a loss of speech. The specialist usually assesses the severity of the disease with the PANSS (Positive And Negative Syndrome Scale) described in 1987 by Dr. R. S. Kay in the United States.

Schizophrenia is present in all latitudes and in all cultures. However, the impact of schizophrenia tends to be highest in the Middle East and East Asia, whereas Australia, Japan, the United States, and most of Europe have typically a low impact. It ranks among the top 10 causes of disability in developed countries worldwide. Twice as prevalent as Alzheimer's disease, it concerns, depending on the country 0.5%–2% of the population, but the variance is difficult to track due to differing measuring standards in many countries. Men are 40% more frequently affected than women, but with a late form occurring around the age of 40–45 years and less common in women. The prognosis of the disease is however better in women because it is less severe and with frequently more positive symptoms responsive to neuroleptic treatments. It has been suggested that this difference according to sex is based on the ability of estrogen to replace vitamin D during the activation of serotonin synthesis (Patrick and Ames 2015).

The approximate number of people in the United States suffering from schizophrenia is more than 2.2 million compared with 6–12 million people in China; 4.3–8.7 million people in India; and 285,000 people in Australia and in Canada (http://www.nimh.nih.gov/index.shtml). In the United States, schizophrenia now costs about US$63 billion per year for direct treatment, societal, and family costs.

In Europe, it was estimated in 2011 that nearly 5 million people suffered with these disorders, compared with 3.7 million in 2005 (Wittchen et al. 2011). In France, nearly 600,000 people would be affected. Patients with schizophrenia represent 20% of full-time psychiatric hospitalizations and 1% of the total health expenditure (data from INSERM). It is important to note that life expectancy of these patients is, on average, 10 years less than that of the general population. Forty percent of the patients attempt suicide, and 10% manage to terminate normally their life.

What is the origin of this disease? For a long time doctors included heredity as a mean to transmit the disease from one generation to the next, but more recently it has been proved that no specific gene is involved. Indeed, it is increasingly clear that these patients have a higher tendency to have rare genetic mutations that involve several genes, in turn disrupting brain development. The onset of schizophrenia is unfortunately unpredictable; however, its hereditary character is underlined by the fact that a person has 10 times more disease risk if a relative (uncle, aunt, grandparent, or cousin) is already affected. Compelling evidence is provided by the observation of identical twins because one will have a 50% risk of being affected by the disease if his or her twin already has it.

In the early 1960s, when neuroleptics were used, an excess of dopamine was formulated. Other neurotransmitter systems have also been mentioned, such as those based on glutamate, GABA, and serotonin. Today, it remains difficult to define the true source of the brain dysfunction generating schizophrenia. Apart from genetic or purely functional explanations, clinicians have also thought of the neurodevelopmental and environmental causes. Thus, psychosocial factors still poorly defined, infections, and malnutrition before or after birth have been invoked.

In 1981, Dr. O. D. Rudin, a supporter of the use of fish oil capsules in the United States, hypothesized that schizophrenia could be linked to a ω-3 fatty acid deficiency (Section 6.1.3.1) (Rudin 1981). D. F. Horrobin explained the effect by the progressive enrichment, for over a century, of our diet with saturated and ω-6 fatty acids, largely replacing ω-3 fatty acids in membrane phospholipids (Horrobin 1998). Being given the main source of ω-3 fatty acids (fish and shellfish), their scarcity in the Western diet is also accompanied by a decrease in the vitamin D supply (Sections 6.1.3.2 and 7.4). After a large survey in nine countries, Danish researchers provided further evidence by showing that the prevalence of schizophrenia is much lower in people eating large amounts of vegetables and seafood (Christensen and Christensen 1988).

6.1.3.1 ω-3 Fatty acids

6.1.3.1.1 Epidemiological investigations Almost 40 years ago, the first mention of a possible link between schizophrenia and a disturbance of the fatty acid metabolism via prostaglandins was introduced by D. F. Horrobin, an English specialist of the relationships between essential fatty acids and various diseases (Horrobin 1977). Subsequently and almost without exception, several studies have shown that ω-3 fatty acids were less abundant in the plasma (Horrobin et al. 1989), red blood cells (Peet et al. 1995), and even brain in patients with schizophrenia (Horrobin et al. 1991). Some studies have validated these results both at the level of red blood cells (Kim et al. 2014) and cerebral cortex (McNamara et al. 2007b).

Despite about 190 studies published since 2000, the origin of this biochemical situation, consequence, or cause of the disease remains unknown. However, a consensus is emerging to support the hypothesis of an increased catabolism or an abnormal metabolism of fatty acids in all tissues, including the nervous system. Nonetheless, some facts remain unexplained as the observation of a restoration of normal DHA levels in red cell of subjects with schizophrenia when treated with neuroleptic drugs (Sethom et al. 2010).

In line with previous work (Christensen and Christensen 1988), conclusions of particular interest were drawn from a large epidemiological survey performed on nearly 33,000 Swedish women aged 30–49 years (Hedelin et al. 2010). These investigators noticed that eating two fish meals per week divided by 2 the risk of developing the disease observed in those consuming no fish. In contrast, at a higher fish consumption (more than five times per week) they observed a higher risk. So far this result remains unexplained, but it could be related to the various contaminants present in the consumed products (heavy metals and pesticides, Section 7.2).

6.1.3.1.2 Intervention studies Well-controlled intervention trials are unfortunately few, but in almost all cases it has been observed that the impact of a treatment with ω-3 fatty acids on the disease development usually involved the appearance of less severe symptoms after a few weeks. One of the first investigations was that of J. E. Mellor from the University of Sheffield, Great Britain, showing that a daily intake of 10 g of MaxEPA (fish oil enriched with DHA and EPA) resulted after 6 weeks in an improvement in schizophrenia symptoms, an improvement correlated with a significant increase in the amount of red blood cell ω-3 fatty acids (Mellor et al. 1995).

More recently, an experiment was conducted in India with 28 patients with an average age of 31 years receiving a conventional pharmacological treatment and ingesting also each day for 4 months 360 mg of EPA and 240 mg of DHA (Arvindakshan et al. 2003). At the end of the treatment, the authors found a reduction in all the characteristic symptoms of the disease, and curiously these changes could be observed yet 4 months after the supplementation while the original fatty acid composition of red blood cells was restored. Also, the fatty acid intake was accompanied by vitamin E (400 IU/day) and vitamin C (500 mg/day).

Early supplementation experiments indicated that an important fish oil intake reduced the severity of the negative symptoms of the disease and the dyskinesia (uncoordinated movements), troubles associated with neuroleptic treatments (Laugharne et al. 1996). In addition, the hypothesis of a close link between ω-3 fatty acids and schizophrenia was supported by a prevention trial of psychotic disorders consisting of a daily ingestion of 0.7 g of EPA and 0.48 g of DHA for 12 weeks (Amminger et al. 2010). At the end of the experiment, the authors found a reduction in symptoms (positive and negative) in young subjects. The most surprising outcome from this experiment is that the beneficial effects were maintained 1 year after the end of the treatment, unlike observations done after a treatment with antipsychotic medication. A review of the scientific literature has shown that these conclusions are shared by many authors, all emphasizing the need of an enrichment of cell membranes with ω-3 fatty acids via dietary intake and for long enough to enable a change in brain composition (Akter et al. 2012).

An experiment carried out in Iran in subjects aged about 32 years and sick for 10 years has clearly shown that a daily intake of 1 g of ω-3 fatty acids for 6 weeks increased the efficiency of conventional antipsychotic treatments (Jamilian et al. 2014). The authors emphasized the usefulness of such a treatment in developing countries given the low cost of fish oil and the absence of adverse side effects.

In patients with a first psychotic episode, an MRI of the brain (hippocampus) enabled to verify that EPA induced changes, confirming its neuroprotective effect. However, it seems that EPA supplementation produced no effect in patients with a well-established schizophrenia. Conversely, several studies emphasized that the treatment is more efficient from the very beginning of the disease (prodromal phase).

This type of dietary treatment of schizophrenia by ω-3 fatty acids is therefore an additional tool in the prevention, especially in the treatment of young people with a clear risk of psychotic disorders.

However, it is unfortunate that the most recent clinical studies have included too few patients and were conducted for too short a time. Nonetheless, the majority of results suggest that a ω-3 fatty acid supplementation would be especially beneficial, and above all, without side effects (Akter et al. 2012). This should encourage clinicians to advise their patients to use ω-3 fatty acids as dietary supplement in addition to their pharmacological treatment to control more effectively the deleterious symptoms of schizophrenia. Highlighting a homogeneous category of patients having both significant negative symptoms and low cellular levels of ω-3 fatty acids must alert also the clinicians on a possible heterogeneity of patients with schizophrenia (Bentsen et al. 2012). The discovery of the presence of subgroups in patients with nervous disorders is not new. Right now, the complexity of these pathologies should encourage clinicians to refine their neuropsychological methods used for subject recruitment.

Although many encouraging results have been reported in scientific journals, it is regrettable that the official organizations or associations dedicated to schizophrenia do not promote the opportunities if not to cure, to at least to alleviate the symptoms of this very debilitating disease. It could be important to suggest to the general medical press and to the media to seize the topic, to analyze the scientific findings, and to report the main outcomes. In this way, many patients could be informed of a simple way to alleviate their neuropsychological disorders. The investment is moderate; side effects are few and safe; and the supplementation with ω-3 fatty acids, purified or as culinary preparations of marine products, are fully compatible with the conventional antipsychotic treatments.

What may be the mechanisms of the effects of ω-3 fatty acids on psychotic disorders? These effects have been attributed, as in many other cases, to changes in membrane fluidity, but also to an interaction with dopaminergic and serotoninergic systems in particular, systems that are conventionally associated with schizophrenia pathophysiology (Patrick and Ames 2015).

As yet, the unconvincing nature of some results has often been considered the result of individuals previously treated with one or more neuroleptic drugs interfering with the ω-3 fatty acid intake. In addition, patients with schizophrenia often have a poor diet, sometimes combined with consumption of alcohol, tobacco, or illicit drugs that may influence the results observed by clinicians. Other better controlled research is needed to improve our knowledge of the mode of therapeutic action of ω-3 fatty acids, a natural therapeutic already perceived as a promising way to treat many other neuropsychiatric diseases. Currently, about 20 major trials in this area are registered at http://www.clinicaltrials.gov.

Already, it may be considered that in subjects at risk of developing or in development of schizophrenia, the results reported above should encourage clinicians to learn from their patients the frequency of consumption of marine products rich in EPA and DHA. If in doubt, it may be possible to determine their blood fatty acid status. If necessary, it is important to advise deficient patients to supplement their diet, so that they meet safely and with no side effects a ω-3 fatty acid intake close to that recommended by medical authorities (Section 7.2).

6.1.3.2 *Vitamin D*

In 1978, a link between vitamin D and schizophrenia was proposed for the first time by R. A. Moskovitz of the Florida University–Gainesville (Moskovitz 1978). This hypothesis was based on the fact that patients with this disease were born more often during winter or early spring and that their mothers had in the third trimester pregnancy vitamin D blood levels as much as three times higher in August than in February. Thus, these observations led naturally to suggest that the greater frequency of the maternal vitamin D deficiency in winter, shortly before birth, could be responsible for an increased risk in young adults of developing schizophrenia.

Unfortunately, this suggestion has found little impact among clinicians, and it took 20 years for Prof. J. McGrath, Director of the Queensland Center for Schizophrenia Research, Brisbane, Australia, to support the hypothesis of a close relationship between low vitamin D levels in the perinatal period and schizophrenia risk (McGrath 1999). Notably, this scientist contributed significantly to the advancement of our knowledge by publishing since 1988 more than 230 scientific articles devoted to that pathology. To support his hypothesis, J. McGrath has grouped all the data available on the excess of schizophrenia cases among those born in winter or spring, when the concentration of circulating vitamin D is at a minimum. Thus, he studied in detail first subjects living in cities compared to those living in the countryside (with more sun exposure, so more vitamin D) and second, subjects with dark skin who had migrated in Nordic countries (low sunshine so common hypovitaminosis). In 2012, a wide survey, conducted at the University of Oxford, UK, by Dr. G. Disanto on nearly 58,000 English subjects, also came to the conclusion that the risk of developing schizophrenia is at a minimum for those born in summer and maximum for those born in winter, peaking in January (Disanto et al. 2012).

Later, Prof. J. McGrath explored in a Finnish population of more than 9000 people the association between vitamin D supplementation during the first year of life and the risk of developing schizophrenia in adulthood (McGrath et al. 2004). This important clinical experiment revealed that a daily supplementation of at least 2000 IU of vitamin D reduced by 77% the risk of schizophrenia compared to subjects receiving a lower dose. It is regrettable that this first trial of a preventive treatment of schizophrenia has not been confirmed by other clinical teams. More investigations are necessary to specify in a preventive approach the most critical period for performing a vitamin D supplementation during brain development.

It seems very likely that the well-known relationships between vitamin D and nerve growth factor are at the basis of brain damage in case of neonatal vitamin D deficiency (Garcion et al. 2002). These troubles may therefore occur throughout life in the form of neuropsychological disorders as those characterizing schizophrenia, but also multiple sclerosis (Section 5.1.2) and possibly autistic disorder (Section 6.1.5.2).

Unfortunately, it is uncertain as yet that the mother hypovitaminosis is predominant in triggering a neuropathological process, as it seems that the vitamin D status in pregnant women has no effect on the risk of developing schizophrenia at an age close to 18 years (Sullivan et al. 2013). In contrast, it may be considered that very low circulating vitamin D levels seem determined by the month of birth. Indeed, in 2015 Dr. G. Lippi has shown that subjects born in winter had significantly lower vitamin D levels than those born during an extended sunshine season (Lippi et al. 2015).

In adult patients with schizophrenia, a vitamin D deficiency has been repeatedly reported by clinicians, despite the variety and the importance of the studied human groups. It has been noticed many times that the severity of symptoms is higher when the level of circulating vitamin D is lower.

To summarize the work published between 2009 and 2013, presenting all the strict criteria, it is necessary to keep in mind the recent meta-analysis of Dr. M. Belvederi Muri, University of Parma, Italy, carried out on a selection of seven published studies involving 523 patients in total compared to 7545 controls (Belvederi Muri et al. 2013). All cited authors, without exception, mentioned a vitamin D deficiency in subjects with schizophrenia compared with healthy subjects. As a matter of prudence, it must be emphasized that this deficiency may not be the cause of the disease, but it may result from an unbalanced food or an insufficient sunlight exposure, both situations often experienced by patients.

The links between vitamin D and schizophrenia remain unknown, but some recent research brings new hope for setting specific therapeutic targets to prevent or even treat a large number of affected subjects. For example, in 2010, the demonstration of a close relationship between geography and genes associated with vitamin D and some neuropsychiatric diseases such as schizophrenia has confirmed at the molecular level the link between psychiatric illness and vitamin D (Amato et al. 2010). Moreover, the authors suggested that schizophrenia may result from a dysfunction of the vitamin D metabolism according to the geographic latitude, thus explaining the greater number of neuropsychiatric diseases in most northern countries. The importance of such a correlation suggests that this mechanism could be involved in other pathologies more or less linked to vitamin D.

More recently, metabolic studies have shown that the links between vitamin D and schizophrenia may result from an increased synthesis of the amino acid proline, likely because of neurotransmission disorders (Clelland et al. 2014). In patients, an exploration of the blood proline levels has shown that many of them present this metabolism imbalance. A pharmacological intervention on these mechanisms could be a means to effectively reduce the disease symptoms even in young adults. An increase of psychiatric disorders such as those characterizing schizophrenia could also be the result of an inhibition of serotonin synthesis linked to vitamin D deficiency (Patrick and Ames 2015).

After finding in animals that vitamin D slows the degeneration of the hippocampus, thus playing a neuroprotective role (Landfield and Cadwallader- Neal 1998), recent anatomical studies using medical imaging (MRI) have reinforced the already compelling links between vitamin D and schizophrenia (Shivakumar et al. 2015). These studies showed for the first time that in young subjects with schizophrenia the volume of their hippocampus was proportional to their vitamin D blood levels, the majority of patients being in a state of vitamin insufficiency or deficiency. Although a causal link between both parameters is not established definitively, these observations suggest strongly that vitamin D deficiency is closely associated with abnormalities of the hippocampus, in relation to the onset of the characteristic symptoms of schizophrenia.

Currently, only five clinical trials in this area are registered (http://www.clinicaltrials.gov).

Much research is still required before deciding the validity of hypothetical relationships between schizophrenia and vitamin D. However, experts stress the need to detect as early as possible any vitamin D deficiency and to correct it as a priority in patients at risk or already diagnosed, although the effectiveness of such intervention in adults is questionable. The situation is different for newborns and pregnant women who must be provided nutritional or supplementation advice to restore, if necessary, a suitable vitamin D level. It is hoped that a preventive administration of vitamin D in pregnant women and newborns, as it is now practiced in France on a large scale, will contribute as it has been observed in Finland to reduce the incidence of some neuropsychiatric conditions, such as schizophrenia, or even neurological diseases, such as multiple sclerosis (Section 5.2.2).

6.1.4 *Attention-deficit hyperactivity disorder*

Psychiatric disorders characterized by an attention deficit with or without hyperactivity disorder, also known as ADHD, are a frequent problem occurring in childhood and persistent in most cases in adulthood. These disorders are classified in the broad category of "disorders of autism spectrum" that also accounts autism, Asperger's syndrome, mental retardation, anxiety, and communication or mood disorders.

Although still poorly understood, ADHD is a genuine neuropsychological disorder and not a psychopathological disorder or a neurological disease. The Diagnostic and Statistical Manual of Mental Disorders (DSM IV-2013) has classified ADHD in the group of the neurodevelopmental disorders. The symptoms affect daily operations and often require an appropriate intervention. In this respect, science is evolving rapidly and several treatment options are available.

The specialists of mental disorders in children and adolescents consider ADHD as a behavioral disorder characterized mainly by three symptoms: difficulty in fixing the attention, motor hyperactivity (hyperkinesia), and high impulsivity (see the National Institute of Mental Health site http://www.nimh.nih.gov/health/topics/attention-deficit-hyperactivity-disorder-adhd/index.shtml). In most subjects, attention deficit and hyperactivity coexist, sometimes inattention or hyperactivity may be predominant, with that distinction creating three subtypes for the same disorder. Thus, the manifestations of ADHD vary from one individual to another, the three symptoms occurring very differently depending on the age or the context of the subject life. In all cases, symptoms interfere with social, academic, or professional activities.

It is a chronic disorder that concerns mainly children of preschool age. The impulsivity often leads to disinhibited or worse aggressive behaviors. In general, support is considered when these symptoms become a handicap for the child. ADHD is often associated with other mental disorders (anger, provocation, aggression, obsessive–compulsive disorders, learning disorders) and without treatment it can lead to many psychological complications. The ADHD present in child persists in 70% when they are adult.

In the field of child psychopathology, this disorder is the most common because its prevalence is 3–5% in western countries, with up to two children per class being so concerned. In 2013, it was estimated that among the worldwide population aged 5–19 years, 129 million had ADHD. Estimates from individual studies have indicated that the global prevalence of ADHD in adults ranges from 1.1% in Australia to 7.3% in France. However, ADHD prevalence data may vary widely between studies due to various factors such as population characteristics; methodological, environmental, and cultural differences; and variability in identification and diagnostic guideline tools, rather than geographical location *per se.*

In 2011, the Centers for Disease Control and Prevention reported that 11% of US school children suffer from ADHD. The percentage of children with an ADHD diagnosis continues to increase, from 7.8% in 2003 to 9.5% in 2007 and to 11.0% in 2011. The annual societal "cost of illness" for ADHD is estimated to be between US$36 and $52 billion.

In the French population, there are about 400,000 children aged from 4 to 19 years and up to 7% of adults with ADHD.

The etiology of this disease is poorly known, although biological and environmental factors are involved undeniably in its emergence. From the functional point of view, it has been proposed that ADHD is consecutive to an impairment of catecholamine-based nerve conduction, but with also a possible intervention of serotonin.

French clinicians are well aware that dietary interventions may sometimes help to prevent a child's problem in that field. In the United States, the strict focus on pharmaceutical treatment of ADHD encourages clinicians to ignore the influence of dietary factors on children's behavior.

From the perspective of the possible influence of nutritional factors, several lipids have been implicated. Among these, ω-3 fatty acids (Section 6.1.4.1) and vitamin D (Section 6.1.4.2) have resulted in some clinical explorations of high interest.

6.1.4.1 ω-3 Fatty acids

Compared with other neuropsychological disorders, ADHD was the subject of the first hypotheses regarding a possible intervention of essential fatty acids (Colquhoun and Bunday 1981). It is noteworthy that the authors discussed their role or those of their metabolites (prostaglandins) a year before the publication of a fundamental work by R. T. Holman, announcing for the first time the efficiency of a supplementation with a ω-3 fatty acid (linolenic acid) in a linolenic acid–deficient person also subject to various neurological abnormalities (Holman et al. 1982).

6.1.4.1.1 Epidemiological investigations The clinical symptoms of excessive thirst (polydipsia), eczema, asthma, and other allergies present in children with ADHD have precociously evoked a situation close to that observed in children with essential fatty acid deficiency. The similarity of these symptoms has been confirmed by several clinicians in finding that they were accompanied by very low blood levels of ω-3 fatty acids. Later, several analytical studies showed that children with ADHD had lower serum ω-3 fatty acid levels than the control healthy children (Stevens et al. 1995, Burgess et al. 2000). In addition, a study has shown that reduced levels of plasma ω-3 fatty acids are associated with low scores in psychometric tests focused on the recognition of facial expressions of emotion (Gow et al. 2013).

Without prejudice to the present mechanisms, some investigations have shown that, in some children or adults, the severity of symptoms was accompanied by a greater essential fatty acid deficiency.

An interesting approach to the problem could be to consider the influence of the nutritional relationships between the mother and her fetus on the risk of developing that disease, as it has been done successfully for behavioral disorders in young children (Steenweg-de Graaff et al. 2015). Fortunately, this large survey (Generation R Study) conducted in Rotterdam, The Netherlands, has shown that high DHA values, measured at midpregnancy, were associated with less behavioral troubles (emotions, anxiety, mood) in 6-year-old children.

6.1.4.1.2 Intervention studies Based on the knowledge gained from epidemiological studies, several clinical trials have been undertaken, including the contribution of ω-3 fatty acids and the assessment of symptoms of the disease by using appropriate and recognized psychological protocols. By contrast with the expectations, the published results are unconvincing because of a too great variability, probably being due to a wide variety of the administered products, the treatment times, and perhaps the existence of poorly defined forms of the disease. The biochemical composition of the administered products seems important as the majority of clinical trials reporting results interpreted as positive were obtained only after administration of Environmental Protection Agency (EPA). Moreover, it was discovered in 2008 that phosphatidylserine, rich in EPA and DHA, was twice as effective in improving scores of visual attention than triacylglycerols of similar ω-3 fatty acid composition, as present in fish oil (Vaisman et al. 2008).

Recent work demonstrated that the administration of DHA, unlike EPA, was able to improve after 4 months the symptoms in a subgroup of children with ADHD aged 7–12 with reading and spelling knowledge difficulties (Milte et al. 2012). The majority of reviews on this subject also highlight that children suffering from symptoms related to ADHD could benefit by supplementation with ω-3 fatty acids, especially those exhibiting difficulties with attention and learning. The relative importance of each long-chain ω-3 fatty acid remains unclear, as does the possible synergistic role of ω-6 fatty acids. Currently, it seems that a combined supplement containing ω-3 fatty acids (EPA plus DHA) and ω-6 fatty acids (linoleic, γ-linolenic, or arachidonic acids) is most effective. Several clinical trials have revealed a reduction of symptoms associated with ADHD, at least in a portion of the groups studied, after administration of a mixture of fish oil and evening primrose (*Oenothera*) oil, with this oil being rich in γ-linolenic acid (18:3 ω-6) (Schuchardt et al. 2012). The dyspraxia symptoms that accompany ADHD tend often to be the first mitigated. These results are as yet inexplicable given that dietary ω-6 fatty acids are known to be commonly very (and even too) high.

A recent placebo-controlled trial, conducted in the Department of Psychiatry, Medical University of Utrecht, The Netherlands, under the direction of Dr. D. J. Bos, has shown that a daily supplementation with equal amounts (650 mg) of DHA and EPA for 16 weeks was capable of improving inattention symptoms in young boys, suffering or not from ADHD (Bos et al. 2015). Despite the limited results of that research, it is interesting that ω-3 fatty acids may also increase the benefit of traditional pharmacological treatments, allowing lower doses of drugs and simultaneously reducing their side effects (Barragán et al. 2014).

The variability of the research results may also be influenced by malnutrition, low socioeconomic level, differences in learning abilities, and certain functional deficits afflicting many subjects with ADHD. In fact, clinicians have noted that benefits of supplementation treatments were higher in children having lower blood levels of essential fatty acids than in properly fed children (Frensham et al. 2012).

Even if there is likely no causal link between these nutritional disturbances and the disease, all experts agree on their aggravating role. An imbalance or a deficiency in essential fatty acids in patients could be a factor modulating the symptoms of the disease, without taking part directly in their determinism; e.g., it could alter the efficiency of any supplementation.

A recent analysis of 13 publications on this subject, selected from 366 contributions (Cochrane Database Syst Rev. 2012 7: CD007986) remained unfavorable for ω-3 fatty acid supplementation in children with ADHD. However, the most recent studies have shown positive results, albeit modest, sometimes in certain patient groups (responders) only, whereas others remain indifferent (Puri and Martins 2014). Thus, several results are found similar to those obtained with traditional treatments, likely proving that the troubles associated with ADHD come from different causes. Despite these discrepancies, a treatment over the long term with ω-3 fatty acids could have in some subjects a beneficial effect, especially in the absence of any side effect and enabling perhaps a better efficiency of other treatments.

The current research interest for ω-3 fatty acids to fight against ADHD seems obvious because 35 studies in this area were listed in 2016 on the official website of the National Institutes of Health (NIH) Clinical Trials (http: //clinicaltrials.gov).

The latest investigations suggest that in the near future, more and more intense research effort with larger numbers of subjects should lead to the proposal of an efficient nutritional therapy for the greatest number of children with ADHD. In case of doubt for an adequate intake of essential fatty acids (Section 7.2), a consultation with a nutrition specialist will enable the estimation of the importance of the deficits and the suggestion of a dietary change before any supplementation.

6.1.4.2 *Vitamin D*

Although knowledge of the functions of vitamin D not related to calcium homeostasis goes back 30 years, its involvement in ADHD pathology is only beginning to be discussed. Following numerous work on brain development disorders (Section 2.2), various neuropsychological diseases such as depression (Section 6.1.1.2), and schizophrenia (Section 6.1.3.2), clinicians have explored actively the possible links between vitamin D and ADHD. The search for these relationships was also motivated by the greater frequency of this pathology among poor, overweight, unmotivated children and those living mainly indoors. These situations are indeed often accompanied by an unbalanced diet and also by a low sun exposure, regardless of the location on the planet.

The first study, conducted in Qatar (Weill Cornell Medical College), comparing 1331 patients to the same number of control subjects, showed clearly that children (between 5 and 18 years old) with ADHD have vitamin D blood levels significantly lower than healthy children (16.6 and 23.5 ng/mL, respectively) (Bener and Kamal 2014). Thus, the former children were in a state of vitamin deficiency, whereas the latter children were only deficient (Section 7.4). That comprehensive study also highlights the close association between the disease and the sociological and adverse medical conditions (e.g., low income, little sun exposure, overload weight). The authors also suggested that a vitamin D supplementation in early life may reduce the incidence of ADHD in children.

A smaller study (60 patients and 30 control subjects) performed in Bolu, Turkey, has confirmed that in children aged from 7 to 18 years there was a clear association between low vitamin D levels and high ADHD incidence (Goksugur et al. 2014).

As for schizophrenia (Section 6.1.3.2), an association between ADHD and the month of birth of the subjects was demonstrated. That relationship seems natural if one assumes the influence of a vitamin D deficiency during pregnancy on child brain development and the risk of subsequently developing the disease. This association has been proven through a recently reported extensive epidemiological survey of approximately 1.75 million people born in New York between 1900 and 2000 (Boland et al. 2015). Indeed, the statistical analysis has shown that the incidence of the disease was significantly higher in subjects born from June to November than in those born from January to May. Thus, these observations emphasize the importance of a proper sun exposure during the second and third trimesters of pregnancy.

Recent work by the team of Dr. M. H. Mossin, University of Southern Denmark, Odense, suggests a protective effect of prenatal vitamin D after finding a novel inverse association between neonatal vitamin levels and ADHD symptoms in toddlers (Mossin et al. 2016). The authors have determined in a population of 1233 mother–child pairs that infants (mean age 2.7 years) tested with a specific Child Behavior Checklist questionnaire had an 11% decrease in ADHD scores per 10 nmol/L increase in blood cord vitamin D. Such tight links between vitamin D and early ADHD symptoms have not been described before and have therefore attracted attention.

Further research is necessary to clarify some causal relationships between the learning difficulties of a child and the mother's vitamin D status, possibly influenced by the month of the birth.

It is regrettable that all these studies have not concurrently paid attention to essential fatty acid status, thereby determining their possible contribution to the reported results. In the future, investigators should focus on this problem and get results quickly after using vitamin D alone or in combination with ω-3 fatty acids for the treatment of ADHD. Preventive measures could also be practiced in young children.

Although these results are encouraging, only one study on the effects of vitamin D on the development of autism in children has been listed in 2016 on the official website of the NIH Clinical Trials (http://clinicaltrials.gov).

6.1.5 Autism

Specialists include autism with Asperger's syndrome in all "pervasive developmental disorders." All these disorders are classified in the broad category of "autism spectrum disorders" that includes also mental retardation, impaired communication, and mood and anxiety troubles (Siksou 2012). Most clinical work does not distinguish between autism and other related disorders; therefore at present, it is impossible to accurately assign a result in one or the other of these disorders.

Autism is a complex mental disorder based primarily on a genetic component. Indeed, among autistics who have an identical twin, this twin is also autistic in nearly 66% of cases. This disorder also seems to depend on environmental influences, so it is a multifactorial pathology. Thus, clinicians have described cases of autism associated with rubella, valproic acid (antiepileptic) treatment, and exposure to thalidomide during pregnancy.

Autism is sometimes considered as a psychosis, with the subject denying any contact and remaining in his or her inner world. It is characterized by inappropriate social interactions, a restricted directory of activities and interests, communication problems, language impairment, and an almost complete ignorance of the environment. Children with Asperger's syndrome have poor social interactions with stereotyped behaviors, but they can have also large capacities of perception, attention, and memory.

It seems that the number of children with autism spectrum disorders is constantly increasing; a US study has determined that in children born in 2002, 1 in 68 was affected by the illness against 1 in 150 for those born in 1992 (Centers for Disease Control and Prevention, http://www.cdc.gov/ncbddd/autism/data.html). The causes and significance of these recent changes are not yet explained (Kim et al. 2011). Among the possible causes, the changing of dietary fatty acid supply and also the recent behavior change in relation to sun exposure, with an obvious impact on the status of vitamin D (Section 6.1.5.2), may be considered.

Globally, the overall prevalence would be around 1.9% (21.7 million people); however, a great disparity exists between countries. The lowest values are registered in several countries of northern Europe, with the highest being in North Korea.

The newest estimate of autism prevalence among the US children remained unchanged, at 1 in 68 from 2010 to 2012.

In 2011 it was estimated that approximately 0.6 million people were suffering from these disorders in Europe (Wittchen et al. 2011).

In France, autism affects about 30,000 children. It grows mainly between 5 and 8 years, with a rate 4 to 5 times higher among boys than girls. Several autism plans were launched by the French State, the last (and third) covers the period from 2013 to 2017 (http://circulaire.legifrance.gouv.en/pdf/2014/07/cir_38551.pdf). Unfortunately, no mention is made to any nutritional intervention in this plan program.

In a child, the existence of autism and the severity of the disorders are evaluated through a battery of tests suitable for preschool age. The best known test is the Childhood Autism Rating Scale (CARS), developed in the United States (Schopler et al. 1980). It uses five behavioral themes rated from 1 to 4 depending on the extent of the troubles.

The structural causes of autism are still unknown; however, recently several studies imply changes in the composition of membrane lipids in nerve tissue of affected individuals and vitamin D status may be important.

6.1.5.1 ω-3 Fatty acids

6.1.5.1.1 Epidemiological investigations In autistics, several clinical signs such as excessive thirst (polydipsia), frequent urine production (polyuria), dullness, and dry and breaking dander have alerted about a possible deficiency in essential fatty acids. The first report of a small concentration of DHA and EPA in the red blood cells of two subjects with autism, compared to two healthy subjects, was done in 2000 (Bell et al. 2000). In addition, two subjects with Asperger's syndrome were not different from controls. In studies on the composition of plasma phospholipids, similar results have been described (Vancassel et al. 2001). Few publications appeared in the following 10 years, but in 2010, specific work by the team of Dr. J. G. Bell, University of Stirling, UK, showed that in 45 autistic children the changes of the fatty acid composition of red blood cells and plasma indicated an unbalance in the essential fatty acids at the expense of ω-3 fatty acids (Bell et al. 2010).

In all cases, the part played by the diet, in addition to other intrinsic metabolic factors, was the object of very little research. A recent Spanish study performed on 105 children with autism has however reported that patients had a lower dietary intake of ω-3 fatty acids, compared to typically developing children (Marí-Bauset et al. 2015). Some authors have formulated the assumption that the frequently found low levels of ω-3 fatty acids could be related to a slowdown of maternal metabolism, when pregnancy occurred in old age or in multiple pregnancies. Prematurity was also cited as a possible cause of autism, given the late fetal accumulation of polyunsaturated fatty acids. The role of the maternal intake of essential fatty acids through breastfeeding seems to be involved if we consider the results obtained by the team of the American neurologist A. Ascherio, Harvard University, Boston, Massachusetts (Lyall et al. 2013). This epidemiological study, one among the most comprehensive, has shown in fact that in a group of more than 17,700 pregnant women, those with the lowest dietary intake of ω-3 fatty acids had an increased risk of giving birth to autistic children. The same was observed in pregnant women with a very low linoleic acid intake, the major ω-6 fatty acid in the Western diet.

6.1.5.1.2 Intervention studies In children with autism, several attempts have been made to discover the curative effects of a ω-fatty acid supplementation. The trials are few and often conducted on small groups of individuals of very different ages, likely explaining the variability of results. Thus, in 2007, improvements in stereotyped behavior and hyperactivity were obtained by G. P. Amminger, Austria, in supplementing some boys daily for 6 weeks with 0.7 g of DHA and 0.8 g of EPA (Amminger et al. 2007). With similar treatments, several research teams have also observed significant improvements in behavior related to the disease, but concerning only a part of the explored children.

Moreover, it seems that the benefits of a ω-3 fatty acid supplementation are no longer observed after a certain age, because no effect could be detected in autistic subjects beyond 18 years.

If the previously reported work gave some hope for an autism treatment with nutritional supplementation, it must be recognized that until 2010 the review of literature did not allow any definitive conclusions on the subject. Some studies detected no significant effect except a slight decrease in hyperactivity in subjects treated with a mixture of EPA and DHA (Bent et al. 2011).

Hope can still be considered since the publication in 2012 of the results concerning an interesting Japanese experiment (Yui et al. 2012). The authors supplemented children who were suffering from Asperger's syndrome daily for 16 weeks with a mixture of arachidonic acid (240 mg) and DHA (240 mg). Although the size of the experimented groups was small (seven treated and six placebo), the results clearly showed that the essential fatty acid supplementation significantly improved the social interactions of the young patients.

Similar results were reported by a Singapore team in 41 young children with "disorders of the autism spectrum" supplemented each day for 12 weeks with 1 g of ω-3 fatty acids (192 mg of EPA and 840 mg of DHA) and arachidonic acid (66 mg) (Ooi et al. 2015). The study of the neuropsychological results by the experts and the opinions of parents revealed an improvement of the scores for awakening, recognition, communication, motivation, and attention. Unfortunately, the authors did not identify with precision the type of autism disorders from which the children were suffering.

These types of studies need to be extended to broader groups that are clinically well defined and as homogeneous as possible (age, sex).

In 2016, two large trials in this area were declared in the WHO International Clinical Trials register (http://apps.who.int/trialsearch/), and eight completed studies are reported on the US site (http://www.clinicaltrials.gov).

In the treatment of autism, as of Asperger's syndrome, research should be better controlled, especially done during longer periods and with more subjects. It is premature to talk about final recommendations on the role of ω-3 fatty acids, and perhaps ω-6 fatty acids. However, it seems important that from the diagnosis of the disease the intake of these nutrients complies in the young patients with standard values set by medical authorities (Section 7.2).

6.1.5.2 Vitamin D

6.1.5.2.1 Epidemiological investigations The effect of the environment on the development of autism was considered soon after epidemiologists noticed a significant prevalence increase in a homogeneous and relatively large population. In several countries, the prevalence of autism was stable until 1980, but then it steadily increased. These changes have been well described around Gothenburg in Sweden where the prevalence was 4 per 10,000 inhabitants in 1980 and 11.5 in 1988, with the increase being globally estimated at about 3.8% per year (Gilberg and Wing 1999). In England, in a population residing south of the Thames, a prevalence of 39 per 10,000 people was observed in 2006, whereas it was only the half 5 years ago in the same region (Baird et al. 2006).

Many hypotheses have been advanced to explain these worrying observations. They usually involved special exposures of pregnant women to environmental factors that could alter the genetic susceptibility for autism (London 2000). The nature of these factors remained obscure until 2008 when Dr. J. J. Cannell showed that the main cause could be a vitamin D deficiency in pregnant women, in very young children, or both (Cannell 2008).

This suggestion based on various clinical outcomes prompted several research teams to explore in detail the vitamin D status in autistic subjects. A review has reported the results of three investigations conducted in

Egypt, United States, and Sweden, with the latter showing clearly that vitamin D deficiency is very common in patients with autism (Kocovska et al. 2012). The study in Sweden on Somali women with children having autism also confirmed that their blood vitamin D level was 30% lower than that of women who had no autistic children, with the sampling being made in the spring, a time corresponding to the lowest vitamin D intakes (Fernell et al. 2010). The same author has recently studied the vitamin D status in 58 siblings of varied ethnic backgrounds, with one child of each suffering from spectrum autism disorder (Fernell et al. 2015). Comparing siblings, it seemed that all the sick children had a blood vitamin D level lower than that of healthy children.

It is remarkable that the dietary intake of vitamin D was regularly considered insufficient in children with autism, probably resulting from a selective behavior toward food (Kocovska et al. 2012). This finding is important because it is well known that autistic children are reluctant to go outdoors, thereby reducing the possibility of skin vitamin biosynthesis. Another interesting situation is that of social groups with low income explored in the United States (Shamberger 2011). Indeed, this study showed that in states where an exclusive breastfeeding of infants was practiced, the prevalence of autism was the highest. Therefore, this observation proves the need for a supplementation in lactating mothers, especially in cases where a vitamin D deficiency was demonstrated.

In connection with the importance of the natural biosynthesis of vitamin D by the skin, a recent review has collected all studies describing in Nordic countries an increased autism risk among dark skin or veiled women (Dealberto 2011). All the results are in favor of the hypothesis of a close association between a maternal vitamin D deficiency and a risk of autism in unborn children. Moreover, the author emphasized the importance of monitoring the vitamin D status in pregnant women, especially among dark skin or veiled immigrants. The problem was expanded through a large survey in the United States made by Dr. W. B. Grant, an expert in this field at the University of San Francisco, California (Grant and Cannell 2013). This author has found that the prevalence of autism disease in children aged 6–17 years was inversely related to the wintertime solar ultraviolet B radiation in various US states, as well as for people with white or black skin.

As for multiple sclerosis (Section 5.2.2), epilepsy (Section 5.3.2), and schizophrenia (Section 6.1.3.2), the prevalence of autism in a population seems to be influenced by the month of birth of patients. That effect is further evidence of a relationship with the capacity of the skin to synthesize vitamin D under the influence of solar radiation. Thus, after analyzing a large number of publications from 11 countries, Dr. W. B. Grant noticed that the period corresponding to the birth of a higher proportion of children with autism is the spring or the summer for middle latitudes and the winter for high latitudes, with these observations corresponding to gestation periods with a minimum of sunshine (Grant and Soles 2009). This effect was found again by the study of Dr. E. Fernell, but only among those born in Sweden or other countries, with the exception of countries in Africa and the Middle East (Fernell et al. 2015). These findings reinforce the hypothesis of the autism development linked to vitamin D deficiency, specifically to its seasonal variations.

In addition, the prevalence of the disease increases with the latitude of the place of patient birth, with that place conditioning consequently the time to sun exposure of subjects and therefore the amplitude of the seasonal variations of the blood vitamin D level. In France, despite a substantial sunshine, this amplitude is important and is not hidden by low dietary vitamin D intakes. As in many European countries, it is possible to conclude that as a result of the scarcity of foods fortified with vitamin D and supplied to consumers, the observed deficiency state arises as a consequence of a lack of political will to encourage a supplementation practice (Ovesen et al. 2003).

The beneficial effects of vitamin D on autism-related disorders are probably due, as for other neurological or psychiatric disorders, to its interaction with dopaminergic and serotoninergic systems in particular, all associated with behavior disturbances (Patrick and Ames 2015).

6.1.5.2.2 Intervention studies What is known about the options of autism treatment with vitamin D? One recently published trial fills the gap. An Egyptian team has reported that the administration of vitamin D (300 IU/kg/day) for 3 months to autistic children deficient in vitamin D significantly improved disorders in about 80% of subjects (Saad et al. 2016). At the end of treatment, the blood vitamin D levels had recovered an acceptable level in all subjects who received the treatment.

Up to now, only one team in Portland, Oregon, had reported promising results concerning a preventive treatment during pregnancy and early childhood (Stubbs et al. 2016). The study was performed prescribing vitamin D at a dose of 5000 IU/day during pregnancy to 20 mothers having already children with autism. The newborn siblings, at high risk for the recurrence of autism, were also prescribed vitamin D (1000 IU/day to the age of 3 years) and followed for 3 years. The final outcome was 5% developed autism in contrast to the recurrence rate of about 20% in the literature for that children population. Notably, this is the first prospective study of the hypothesis of using vitamin D in an attempt to prevent the recurrence of autism in a high-risk group of newborn siblings. Although difficult to conduct, more research with larger numbers and a control group are necessary to enable precise recommendations.

A literature review has recently identified a large number of observational studies, but very few intervention trials, investigating the relationships between vitamin D and autism (Mazahery et al. 2016). The authors concluded that despite inconsistent results, there are clear indications that early exposure to inadequate vitamin D may contribute to the etiology of autism. Thus, vitamin D deficiency is highly prevalent in populations with autism, and intervention with vitamin D might be beneficial in reducing the symptoms. Therefore, there is an urgent need for randomized controlled trials of vitamin D in populations genetically predisposed to autism as well as in adults suffering from the disease.

It is to be hoped that the increasingly common practice of vitamin D supplementation in lactating women and newborns, combined with a recommendation of extended breastfeeding, will help to decrease autism prevalence in the population. In 2016, four studies evaluating the effect of vitamin D on clinical outcome in autistic children were reported recruiting or completed on the US site (http://www.clinicaltrials.gov), and two studies were declared on the WHO International Register of clinical trials (http://apps.who.int/trialsearch/).

Parents are strongly recommended to encourage their children participate in outdoor activities to obtain the maximum benefit of skin synthesis of vitamin D, while avoiding excessive sun exposure. For many authors, the observed increase in recent years of the number of subjects with autism in all countries is due to a change in children's lifestyles, with decreased sunshine exposure following the increase of activities practiced indoors (television, computers). In addition, it is imperative to recommend pregnant women to monitor their blood vitamin D levels to reduce the risk of developing autism in their children. These children should also benefit if necessary from a prolonged vitamin D supplementation.

6.2 Other personality and behavior disorders

It is accepted that subjects with attention-deficit disorder may have also, sooner or later, behavioral problems characterized by impulsiveness often accompanied by aggression and violence ("conduct disorder"). The subjects do not recognize the rights of others; their communication is restricted to physical or verbal aggression and often the subjects become cruel and destructive, or perform acts of thievery. In young children, when these disorders appear, they are socially disruptive and may persist among adults who become predisposed to injure others (homicide and imprisonment) or themselves (suicide attempts).

The estimate of the extent of personality and behavior disorders described in this chapter is most often performed using a scale called Hopkins Symptom Checklist 90 (SCL-90) (Derogatis et al. 1973). The test as a questionnaire is designed primarily to assess in patients over 13 years various psychological and psychiatric problems during pharmacological treatments. It comprises 90 questions, rated 0 or 1, divided into nine themes: somatization, obsessive–compulsive disorder, interpersonal sensitivity, depression, anxiety, hostility, phobia, paranoid ideation, and psychoticism. The sum of the scores for each theme gives a graph enabling to establish a personal profile and to follow its evolution throughout treatment.

Another very commonly used scale, Brown-Goodwin Assessment (BGA), was established in 1979 to help military enrollment (Brown et al. 1979). The scale is established by a professional on the basis of a direct interview of candidates or patients. Eleven themes are analyzed and rated from 0 to 4 according to the seriousness of the facts. These themes are anger, fight, aggression, behavior at school, civic discipline, antisocial behavior without intervention of police, antisocial behavior with police intervention, military discipline, military discipline with judicial system, damage to objects, and verbal aggression.

6.2.1 *Aggressive behavior*

Psychologists explain aggression as a desire for domination with malice leading to ensure the subject that his or her own power will be recognized and that could drift toward sadism. Psychiatrists propose to find at the base a concept of frustration, a theory of Freudian inspiration, but the reality of a socially learned behavior cannot be eliminated. There is nothing to oppose these two theories. Aggressive behavior may be characterized by verbal attacks, physical attacks, or both. Aggressiveness generates most often a violence that, in children as in adults, may start with intimidation and harassment and then progress to more serious forms such as theft, sexual assault, hold-up, and even murder. In adults, it is now considered that an aggression that does not meet a real threat is a sign of mental disorder, a comorbidity accompanying frequently schizophrenia or alcoholism.

Psychologists have established that aggression stems from a form of rejection of other people and also of oneself and of refusal to accept reality. The hostility facing others remains complex to understand because it manifests variously; therefore, its assessment requires several psychological tests adapted to the specific conditions of each country (Section 7.12). For about 30 years, much research in this field has highlighted several biological factors related to impulsiveness and violent behaviors. Among these determinants, clinicians have found that abnormally low levels of a serotonin derivative, 5-hydroxyindoleacetic acid, were present in violent military (Brown et al. 1979), impulsive criminals (Linnolila et al. 1983), and even arsonists (Virkkunen et al. 1987a). Similar results were obtained from persons categorized as violent and with suicidal behavior (Asberg et al. 1976).

Following this research, investigations were extended to lipids influencing likely brain functions by disrupting serotonin metabolism. Among these lipids, ω-3 fatty acids (Section 6.2.1.1) and vitamin D (Section 6.2.1.2), already involved in various other neuropsychological disorders, have been the subject of most research efforts. Cholesterol also has been investigated regarding violence (Section 6.2.1.3).

6.2.1.1 ω-3 Fatty acids

It is accepted in everyday life by the medical profession that an angry or hostile behavior greatly increases the risk of developing heart troubles that can be fatal. Conversely, a treatment leading to reduce hostility tends to limit the risks of heart attack (cardiac ischemia) (Williams and Littman 1996). With the established relationships between ω-3 fatty acids and cardiovascular diseases (Leray 2015), it seems obvious to explore whether these dietary lipids may also improve impulsivity and the hostility state even without any cardiovascular impairment.

6.2.1.1.1 Epidemiological investigations Early research in this area has demonstrated the presence of very low levels of plasma ω-3 fatty acids in violent and impulsive individuals. Dr. M. E. Virkkunen, University of Helsinki, Finland, was the first to show that usually violent men and criminals (murderers, arsonists) had very low blood levels of ω-3 fatty acids, mainly DHA, whereas ω-6 fatty acids were more concentrated than in normal subjects (Virkkunen et al. 1987b). So, an impaired blood composition focusing on essential fatty acids is revealed because it has been reported in subjects with ADHD (Section 6.1.4.1). This similarity demonstrates the existing relationships between attention-deficit disorder with hyperactivity and behavior troubles oriented toward aggression, with the latter likely belonging to ADHD-associated disorders. Indeed, we know that among these associated impairments, oppositional disorder with provocation and conduct disorder with aggressiveness may be present in 25%–66% of cases, especially if the ADHD treatment was late (Szatmari et al. 1989).

These biological characteristics were found also in young violent offenders (Corrigan et al. 1994) and among boys aged 6–12 years, regularly angry with sleep and learning problems (Stevens et al. 1996). Low levels of ω-3 and even some ω-6 fatty acids have also been observed in the blood of regular cocaine consumers with aggressive behavior (Buydens-Branchey et al. 2003a). More recently, it has been shown that the plasma EPA concentration was the biochemical indicator the most highly correlated (negatively) with aggression and impulsivity in young subjects, but only in addicts to alcohol, cannabis, or cocaine, even if in remission (Beier et al. 2014). This relationship was found to not originate from an inadequate EPA dietary intake.

The knowledge of the relationships between the ω-3 fatty acid status and drug addiction is just getting started. Indeed, experiments in animals have emphasized the role of these lipids in the abuse of some substances controlling serotonergic and dopaminergic systems. The first exploration of that complex problem was reported in 2003 by Dr. J. R. Hibbeln, an American expert on these issues. He showed in 32 patients consuming cocaine that the lower the plasma level of the ω-3 (and ω-6) fatty acids, the more elevated was the risk of relapse in the short term (Buydens-Branchey et al. 2003b). The authors insisted that low levels of these fatty acids at the beginning of the treatment were the best predictor of a future relapse. Unfortunately, no direct causal link has been discovered between the consumption of seafood (fish, mollusc, shellfish) and drug use, such as cannabis or cocaine. It is amazing that no large-scale clinical study has been undertaken on this subject, especially because this hypothesis was also raised for smoking (Zaparoli and Galduróz 2012) and alcoholism (Le-Niculescu et al. 2011).

It is obviously desirable that research should be a starting point for explorations extended in all addiction areas. Moreover, potential treatments with essential fatty acids are known to be well tolerated and inexpensive, in contrast with the expenses currently incurred in the fight against these behavioral disorders.

To explore the involvement of neurotransmitters and fatty acids in the brain function, Dr. J. R. Hibbeln studied men guilty of domestic violence (Hibbeln et al. 2004b). He demonstrated an inverse relationship between plasma DHA levels and the concentrations of corticotropin-releasing factor, an hypothalamic hormone known to be involved in the responses to stress as violent and defensive behaviors.

Because there could be interference of ω-6 fatty acids in the physiological process underlying an aggressive behavior, a cooperative study of Australian researchers recently published the values of the "ω-3 index" (Section 7.1) of red blood cells measured in 136 adult prisoners with aggression troubles and attention deficit (Meyer et al. 2015). The results have shown that all the scores on the parameters defining the subject behavior were negatively correlated with the ω-3 index values. Despite great variability, it is clear from that study than those having the lowest ω-3 index were the most aggressive and had the most severe attention disorders. The interest of this work is to have taken into account the ω-6 fatty acid levels, knowing that their abundance in the tissue influences the ω-3 fatty acid requirements.

Further research was conducted by Dr. A. Zaalberg, Altrecht Institute of Psychiatry, Den Dolder, The Netherlands, on 51patients interned after physical aggression (Zaalberg 2015). The fatty acid composition of red blood cells was determined, and neuropsychological tests were performed to estimate the intensity of the psychopathological symptoms by using the General Health Questionnaire (GHQ-28) (Pariente and Smith 1990) and especially aggressiveness by using the Social Dysfunction and Aggression Scale (SDAS) and the Aggression questionnaire (Meesters et al. 1996) (Section 7.12). The authors reached the conclusion that there was a close association between DHA levels and total ω-3 fatty acid concentrations, the ω-3 index, and aggressiveness (Figure 6.2). Besides the small amounts of ω-3 fatty acids, the authors also measured very low vitamin D levels, at least for the majority of the subjects studied. Among the 51 subjects, 32 had insufficient concentrations and only 8 had values above the national standard (28.5 ng/mL [75 nmol/L]). However, no correlation could be detected between vitamin D levels and the intensity of aggressive events.

That rigorous work combining fatty acids and vitamin D underlines the need to take into account several biochemical parameters in biological samples. Obviously, the characterization of the effects strictly related to a specific substance could be established only with the help of supplementation experiments, with any synergy between several substances being obviously not ruled out.

A confirmation of the involvement of ω-3 fatty acids in human aggressive behavior may also be obtained by exploring among populations, or in a given population, a possible association between these behavioral disorders and fish consumption, the main source of ω-3 fatty acids.

The first major "ecological" study of the relationships between fish consumption and mortality by homicide (not suicide) in 26 countries was reported by J. R. Hilbbeln in 2000 at the 4th Congress of the

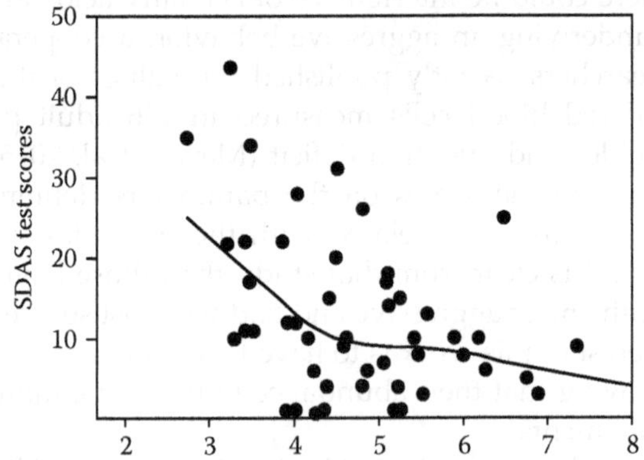

Figure 6.2 Relationship between SDAS scores and ω-3 index. (From Zaalberg, A., Nutrition, neurotoxicants & aggressive behaviour, PhD Diss., Nijmegen, 2015.)

International Society for the Study of Fatty Acids and Lipids (Hibbeln 2001). Although estimating the number of homicides is an expression of violence in extreme situations, it remains a simple and precise method. The author noted for the year 1995 homicide rates in the WHO statistics and fish consumption in those of FAO. He dismissed the United States from his list of countries because at that time the homicide rate was very high (20 per 10,000), likely influenced by the ease of buying arms and by the violence displayed in media. Taking into account the extreme values, the homicide rates between countries ranged from nearly 10 times and fish consumption about 15 times. Hibbeln has determined a decreasing logarithmic relationship highly significant between these two parameters. Furthermore, he estimated that below a consumption threshold of 10–20 kg of fish per year, the homicide rate increased very quickly. That correlation is therefore consistent with the previous results based on blood biochemical parameters. That communication was released to the media (*New York Times*) 5 years later (April 16, 2006) as an article titled, "Does eating salmon lowers the murder rate?" It was widely discussed in the US media and perhaps it contributed to the awareness of the population focusing on the interest of increasing the consumption of seafood products, while decreasing that of oils and any product rich in ω-6 fatty acids. This hypothesis was also ascertained 3 years later by J. R. Hibbeln by using statistics from five countries. These recommendations may be also delivered to the French population because all official surveys have shown that ω-3 fatty acid intakes are largely insufficient because they represent only one half of the dietary reference intakes advised by medical authorities (Leray 2015).

Following these first quite spectacular results that were well reflected in the media, several epidemiological studies attempted to explore more broadly the association between dietary intake of ω-3 fatty acids and violent behaviors. A great survey was conducted in the United States with 3581 individuals of both sexes enrolled in three cities (Birmingham, Chicago, and Minneapolis), within the context of the CARDIA study (Iribarren et al. 2004). Analysis of the fish consumption and the estimation of the hostile behavior of subjects aged 18–30 years showed a highly significant relationship between an increase in DHA intake and a decrease in the violence level. Although the authors could not perform blood tests, the results strengthened the hypothesis of a reduced hostility in individuals consuming relatively high amounts of marine fish.

6.2.1.1.2 Intervention studies Naturally, all these studies are not a definitive proof of a beneficial effect on hostile behavior exclusively due to a diet rich in marine fish; indeed, the study of ω-3 fatty acid supplementations is the most effective way to clarify and deepen the subject.

The first experimental study was done in Japan with 41 students for a period of 3 months before a stressful exam (Hamazaki et al. 1996). It turned out that students ingesting daily 3 g of fish oil (1.5 g of DHA and 0.2 g of EPA) maintained a stable behavior throughout the trial, whereas control subjects ingesting soybean oil became much more hostile at the end of the study. Later, the author was able to confirm that a similar treatment was also effective in employees of approximately 50 years old subjected to stressful situations such as videos of real crimes and accidents caused by guilty negligence (Hamazaki et al. 2001).

Obviously, we have to be careful before drawing general and final conclusions because these studies have shown that the education level could also influence the results; thus, a higher education level would induce more significant answers after ingesting DHA.

With this experiment, one can measure the interest of a natural preventive treatment with fish oil during stress periods, such as those experienced before competition or examination. By this means, the decrease in aggressiveness, therefore anxiety, can be surely a pleasant training for psychologically difficult periods, certainly more efficient than many usually used pharmacological treatments.

In 2008, Prof. T. Hamazaki, University of Toyama, Japan, has reviewed several previous work related to the influence of ω-3 fatty acids on various manifestations of aggression (opposition, violence, anger, and aggression) or its opposite friendliness (Hamazaki and Hamazaki 2008). Of the 14 listed studies, 13 confirmed to varying degrees the outcomes set out above, with only one reporting no modification.

Children have been also the target of this research theme, probably due to the finding of increasing adaptation difficulties with manifestations of violence, both in society and in academics. It must be noticed that the reason for aggressive behaviors is the cause of almost half of the pediatric psychiatry consultations. These troubles result in learning difficulties that naturally lead to employment problems, social isolation, violence, and even crime or suicide.

Physiologically, the period of adolescence is one of the times when the prefrontal cortex undergoes maturation involving a growth of neuronal dendrites and an increase in the white matter volume. Furthermore, the executive functions involving attention, emotion, and impulse control are localized in the prefrontal brain area.

It can be considered that the study done by Dr. M. Itomura, Toyama Medical University, Japan, is the first attempt to highlight a reduction in aggressive behavior in young children between 9 and 12 years old (Itomura et al. 2005). In this work, a group of 90 children were fed for 3 months with food (bread, sausage, pasta) supplemented with fish oil so as to provide a daily intake of 0.51 g of DHA and 0.12 g of EPA; the control group was supplemented with vegetable oil. The author reported beneficial results for impulsiveness, but only in young girls. However, the test results were not correlated with the concentrations of blood ω-3 fatty acids. Unfortunately, the difference in response between boys and girls still remains inexplicable, but it reminds us to remain wary of results according to sex in studies on the lipid effects.

In the United States, a similar study at The University of Pennsylvania, Philadelphia, on children from 8 to 16 years old reached positive conclusions, but for both sexes (Raine et al. 2015). The origin of that difference lies likely in a greater treatment time (6 months instead of 3months) and a more important ω-3 fatty acid intake (1 g instead of 0.6 g/day). The author reported that the beneficial effect on behavior was maintained 6 months after the end of treatment.

A recent study conducted in Australia (Mater Health Services, South Brisbane) investigated the effect of a daily supplementation of 4 g of fish oil in children 7–14 years old, selected for their particularly aggressive behavior. Despite the use of many neuropsychological tests, this study failed to highlight a behavior change after 6 weeks, except for a decrease of hyperactivity (Dean et al. 2014). As the author suggested, the treatment time may have been too short for obtaining significant results.

In young adults in good health and without aggressive behavior, a daily supplementation for 12 weeks with fish oil (672 mg of DHA and 100 mg of EPA) decreased impulsivity and aggressiveness, both measured using a battery of specific neuropsychological tests (Long and Benton 2013).

Prison seems to be the ideal environment to highlight a possible effect of dietary lipids on antisocial behavior and violence. Two trials were undertaken and provided substantially similar results. The first experience of supplementation with ω-3 and ω-6 fatty acids performed as a double-blind against control trial was undertaken in an English prison to reduce disciplinary incidents (Gesch et al. 2002). The test subjects ingested a capsule containing 80 mg of EPA, 44 mg of DHA, and 1.4 g of ω-6 fatty acids daily, with the control subjects ingesting a capsule containing only vegetable oil. All received a multivitamin capsule. The authors determined that disciplinary complaints (especially for serious incidents) were reduced by 35.1% after 2 weeks and 26.3% after a 5-month treatment, thus seeming to validate the effect of the fatty acid supplementation on the antisocial behavior of incarcerated subjects.

A very similar test was done in The Netherlands on 220 prisoners with the mean age being 21 years (Zaalberg et al. 2010). The daily double-blind supplementation in half of inmates randomly selected consisted mainly of 0.4 g of EPA and 0.4 g of DHA together with 0.6 g of linoleic acid and 0.1 g of γ-linolenic acid; the other half of the subjects received a placebo. The authors have found a significant decrease of incidents between incarcerated subjects and the prison institutional staff, even neglecting the incidents under the influence of alcohol or illegal drugs. In contrast, neuropsychological tests did not reveal any difference between the placebo and the test group. These results are surprising, but they suggest that it would be necessary to extend these trials to different categories of supervised offenders to improve at little extra expense their living conditions and those of prison staff.

From these studies, it is unfortunately difficult to attribute the profit of treatments only to ω-3 fatty acids because the used preparations contained also ω-6 fatty acids, vitamins, and other trace elements. It is possible that the restoration of other biological parameters were involved in the EPA and DHA effects. In addition, the groups of subjects studied had sometimes more or less defined disorders.

It seems desirable that psychiatric clinics, as prisons, could be the place for experimentation involved in improving the welfare of detained patients or inmates and possibly decreasing the aggression toward others or themselves. Unfortunately, no establishment has currently embarked on this path. Clinical trials should also be programmed to test the individual effects of EPA or DHA at various doses and independently of other substances. The possible profits of these inexpensive tests could lead rapidly to an important economic and social progress.

It seems increasingly clear that research on human behavior cannot be carried out like those on drugs used to fight, as an example, against hypertension. One or even several biological indicators may not be enough, but owing to animal studies several ways involving fish oil in regulating human aggression are clearly emerging. As with cardiovascular diseases or cancers, the data already acquired allow to hypothesize that many cases of violence are caused by the excessive dietary importance of vegetable oils rich in linoleic acid. It is therefore necessary that the relationships between dietary lipids and the aggression mechanisms could be quickly identified. Thus, clinicians will be able to propose simple preventive treatments against violent behaviors that are affecting our societies in remaining a heavy financial burden for all countries. As mentioned by Dr. J. R. Hibbeln, could it be wise to return to millennia practices of Christian churches and Chinese sages who very early associated fish with calm and peaceful behaviors? (Hibbeln 2007).

In 2016, only six trials in this area were declared in the United States (http://www.clinicaltrials.gov).

6.2.1.2 Vitamin D

The relationships between vitamin D and aggressive behavior have been very rarely explored. This temporary statement is not comprehensible if one recalls many results proving an unequivocal association between vitamin D and often-violent situations such as schizophrenia (Section 6.1.3.2) and attention-deficit disorder with hyperactivity (Section 6.1.4.2).

A recent work reported earlier about ω-3 fatty acid effect (Section 6.2.1.1) has also shown that in 51 patients interned as a result of physical attacks, 32 had insufficient vitamin D levels, whereas only eight had levels above the national standard (Zaalberg 2015). However, no correlation has been found between vitamin D levels and the intensity of aggressive behavior as measured by several neuropsychological tests. A similar vitamin D status has been also reported for long-term inmates in Phoenix, Arizona, where 90% of the correctional population had serum vitamin D levels lower than 20 ng/mL (Jacobs and Mullany 2015).

An attempt to explore this type of neuropsychological relationships was made when studying a population of 1095 Iranian teenagers (Ataie-Jafari et al. 2015). The authors have detected that 40% of subjects were vitamin D deficient (concentration <10 ng/mL), 39% were insufficient (concentration between 10 and 30 ng/mL), and only 21% had normal values (concentration >30 ng/mL). The clues of psychiatric distress (anger, anxiety, insomnia, fear, and depression) were closely related to the serum vitamin concentrations. The intensity of disorders was almost twice as elevated in deficient subjects compared with those who were not deficient. However, the violent behaviors were not influenced by the vitamin D status. Note that the assessment of violence was established in using only individual self-reported statements on the frequency of physical assaults and several psychiatric distress situations. The large number of subjects interviewed for that research cannot compensate for the simplicity of the retained tests. Unfortunately, no data on dietary or circulating essential fatty acids have been produced.

It is hoped that further studies will explore this area of violence not well known in prison by using blood analysis of vitamin D and also of ω-3 fatty acids, with the latter being already widely suspected to influence the generation of violent behaviors. Supplementation trials should update these investigations to restore if necessary appropriate lipid intakes, an unavoidable action facilitated by standardized diets and adequate sun exposure.

To date, no clinical trial concerning this specific area has been declared in the international registers.

It is obviously premature to conclude that there is a causal link between psychiatric distress and vitamin D, but the analysis of social data allow to consider a possible influence of the diet, as for sun exposure. In the current state of knowledge and medical practice, the authors of all work in that field may recommend only a better control of vitamin D deficiency mainly in adolescents.

6.2.1.3 Cholesterol

6.2.1.3.1 Epidemiological investigations Several early work suggested that a low blood cholesterol concentration could be associated with a decreased incidence of cardiovascular diseases, but also with the presence of violent behavior often followed by death.

In 1979, the psychiatrist specialist of these questions Dr. M. Virkkunen, Helsinki University, Finland, showed that subjects with antisocial behaviors had usually low cholesterol levels. Later, he explained that among adolescents with attention deficit and hyperactivity disorders, those with an aggressive behavior had lower cholesterol levels than other children (Virkkunen and Penttinen 1984).

The problem therefore seems complex and shows that several nutritional or environmental factors may interfere with neuropsychological parameters. A clarification of behavioral parameters seems also necessary because different types of hostility have been linked or not with serum cholesterol (Hillbrand et al. 1995).

To illustrate this field, the recent publication by the team of Prof. M. Virkkunen (Repo-Tiihonen et al. 2002) brings some answers. That work demonstrated that in a group of 250 criminal offenders, the subjects with a cholesterolemia below the median value and classified as violent (guilty of armed robbery or murder) were seven times more likely to die before the median age of death in the cohort studied. In contrast, those classified nonviolent (arsonists, thieves, fraudsters, and drunken drivers) had a risk to die of unnatural causes eight times higher than other criminals. The mean total cholesterol level of these offenders with antisocial personality disorders was lower than that of the general Finnish male population. Considering the social history of offenders, the authors hypothesized that the circulating cholesterol could be a biological marker in children with antisocial behaviors, with a prognosis value for a further development of criminal behavior.

It would be interesting to establish the possible influence on this type of behavior for the past 20 years of the replacement of cholesterol-rich breast milk by cholesterol-poor infant formula (containing only vegetable lipids). That progressive change in infant feeding might be able to influence the cholesterol status in adolescents and even adults (Horta and Victora 2013).

A large health study was done in Sweden comparing the police records and blood cholesterol levels in 100 criminals among nearly 80,000 people followed for about 30 years. That survey has thus shown that when cholesterolemia decreased, the proportion of criminals in the population increased (Golomb et al. 2000). So, for a cholesterolemia between 2.51 and 2.79 g/L, there was 0.85 violent criminal per 1000 individuals; from 2.26 to 2.50 g/L, the ratio became 1.55 and below 2.26 g/L it reached 2.0.

The authors emphasized that there was no significant causal link between cholesterol and violence, but they noted that cholesterolemia was determined long before the appearance of criminal offenses. It must be recalled that these cholesterol levels are well above the currently recommended levels (1.50–2.00 g/L). Among patients with schizophrenia, no significant relationship could be detected between cholesterol and violence (Steinert et al. 1999).

The study of populations in good mental health may also help to detect the possible predispositions to violence for a precise range of cholesterolemia (Pozzi et al. 2003). This has been done with a large survey of more than 2000 Italian young men aged 19–31 years. That study revealed an association between cholesterol and impulsivity, although less obvious than in subjects suffering from psychiatric disorders but still verifiable in individuals with the lowest cholesterol levels (<1.40 g/L).

6.2.1.3.2 Intervention studies Intervention studies are infrequent and difficult to interpret. During a long-term trial, 149 people were selected in a normal population and were subjected to a new diet with less lipids but rich in complex carbohydrates to improve their cholesterol levels (Weidner et al. 1992). After a 5-year treatment, the authors found low blood cholesterol levels accompanied by an improvement of the emotional state and a decrease of the subject aggressiveness. Note that before intervention, they did not suffer from psychiatric diseases and that no blood sampling or nutritional survey had been carried out. It seems therefore prudent to eliminate any influence of cholesterol during important dietary changes that could alter other biochemical indicators. Thus, in the absence of direct supplementation with cholesterol, ethically impossible to perform, it seems difficult to conclude that cholesterol plays a role in aggressiveness with only the help of epidemiological studies.

To reduce the incidence of cardiovascular diseases, the increasing use of statins to fight against elevated cholesterol levels has prompted clinicians to search for the possible influence of these drugs on all causes of mortality. Could this type of pharmacological intervention help us in understanding that complex problem?

In 1990, Prof. M. F. Muldoon, University of Pittsburgh, Pennsylvania, showed that a cholesterol-lowering diet or a statin treatment decreased, as expected, the *mortality* from acute myocardial *infarction* (−15%) but increased the nonillness mortality (accidents, violence, and suicide) (Muldoon et al. 1990). Ten years later, the same author realized an analytical review of 19 similar studies and found a lack of relationship between reducing cholesterol and nonillness mortality. Nevertheless, a trend toward increased deaths from violence (and suicide) was observed in trials of dietary interventions and nonstatin drugs (Muldoon et al. 2001). Despite some reservations, these results suggest an association between the circulating cholesterol concentration and certain behavioral predispositions oriented to violence and followed by accidents. Similar conclusions were obtained from cholesterol supplementation experiments in Cynomolgus monkeys (Kaplan et al. 1996).

Much work is still required before clear links between cholesterol and aggressiveness can be established. Thus, it seems unavoidable now to pay attention to dietary essential fatty acids or vitamin D. Yet, it seems very important to determine the minimum threshold of blood cholesterol to avoid any impact on social behavior. Such research should help to reassure people who raise questions on the merits of the anticholesterol treatments based mainly on statins.

As with other neuropsychological disorders, the development of aggressive behavior under the assumed effect of low cholesterol concentrations is probably related to the decreased activity of serotonergic neurons. This effect is likely a consequence of a slowed transmission or a decreased activity of specific receptors. Professor B. Wallner, University of Vienna, Austria, has suggested that a reduction of the cerebral cholesterol biosynthesis induced by food scarcity could be able, in prehistoric humans, to start a beneficial aggressiveness for foraging and hunting (Wallner and Machatschke 2009).

6.2.2 *Suicidal behavior*

Suicide is an exacerbated expression of a feeling of violence turned toward oneself. In 1897, Émile Durkheim, the French sociologist and anthropologist, considered to be one of the founders of modern sociology, defined suicide as "the end of life resulting directly or indirectly from a positive or negative act of the victim who knows he is going to kill himself." Willful, deliberate, or not, the act to end one's own life is achieved with violent (e.g., hanging, drowning, firearm, cold weapon) or nonviolent (e.g., poisoning, suffocation, drug overdose) methods, the psychiatric importance of which has not been extensively studied. Moreover, the methods of suicide vary across countries and cultures, with firearms being preferred in the United States, hanging in Eastern Europe, and poisoning by pesticides in China. In the majority (about 90%) of the cases analyzed by specialists in Western countries, and in contrast with China, the suicide attempt or the suicide itself is related to a mental disorder such as depression, bipolar disorder, or schizophrenia. It is often an act of desperation consecutive to alcohol or illicit drug use, drug-seeking behavior being frequently considered as suicidal equivalents. Suicide is rarely motivated by religious or moral considerations. The diversity of situations complicates the understanding of the act itself and contributes certainly to increase the dispersion of biochemical and neuropsychological analyses used by investigators.

From a semantic point of view, there is the suicidal act, grouping suicide and suicide attempt, and also the suicidal thoughts, including all thoughts or beliefs of a person about the end of his or her life.

It is accepted that about 90% of persons dying by suicide previously suffered from a mental disorder, with mood disorders being the most common of the associated diseases. The psychiatric studies have shown that 25%–50% of patients with a bipolar disorder complete at least one suicide attempt and 15%–20% die. The second risk condition is depression, often associated with alcoholism and addiction. Note that 10%–13% of patients with schizophrenia die by suicide, often referred to as psychotic or delirious suicide.

Worldwide, the WHO has estimated that about 1 million people commit suicide each year, thus more than the total deaths caused by wars and homicides.

In the United States, the annual age-adjusted suicide rate is 12.9 per 100,000 individuals; this rate corresponds to an average of 117 suicides per day. Men die by suicide 3.5 times more often than women. In 2014, the highest US suicide rate (14.7) was among whites and the second highest rate (10.9) was among American Indians and Alaska natives. Lower rates were found among Hispanics (6.3), Asians and Pacific Islanders (5.9), and blacks (5.5). It has been estimated that the numbers could be higher (data from the American Foundation for Suicide Prevention).

In Europe (27 countries), the average rate of death by suicide is estimated at 10.2 per 100,000 people. Suicide is thus a serious public health problem even in France where it is responsible for about 11,000 deaths per year (16 per 100,000 inhabitants). This rate makes France one of the five countries of the European Union with the highest suicidal mortality. In addition, the French Health Ministry has estimated that attempts are probably more than 15 times greater than death by suicide. It is the leading cause of death for subjects aged 25–34 years. Higher rates are even noticed in some countries, with the highest being in Russia and Lithuania (>30 per 100,000 inhabitants), Finland (28 per 100,000 inhabitants), and Japan (26 per 100,000 inhabitants). Men have, except in China, higher suicide rates than women.

Worldwide, it has been established that the importance of suicides in premature mortality has more than doubled in the past 30 years. It is therefore understandable that the search of risk factors of suicide is the subject of a major effort by psychiatrists.

Although at present, no scientific evidence allows to predict a suicide gesture, clinicians may use more than 30 tests to evaluate risk assessments, with one of the oldest and most used being the SSI (Scale of Suicide Ideation) (Beck et al. 1979). Undoubtedly, this plurality is the reflection of the difficulty encountered by every professional in the evaluation of a suicidal potential in a patient.

6.2.2.1 ω-3 Fatty acids

Very early, suicidal behavior has been associated with low levels of the serotonin metabolite 5-hydoxyindolacetic acid (5-HIAA) in cerebrospinal fluid, with this status being due to lower serotonin amounts in the central nervous system (Roy et al. 1987).

In addition, serotonin is known to be involved in various psychiatric disorders such as stress, anxiety, phobia, and depression; it is the target of some drugs used to treat these diseases. Low 5-HIAA levels were also associated with reduced levels of ω-3 fatty acids, in particular DHA (Hibbeln et al. 1998). Such levels were much lower in the very impulsive subjects with suicidal thoughts than in nonimpulsive or in control subjects (Cremniter et al. 1999).

An original observation was done in the Ghent University Hospital, Belgium, concerning a close synchronism between the seasonal variations of the blood ω-3 and ω-6 fatty acids, serotonergic markers, and the rate of violent suicides (De Vriese et al. 2004).

All these observations are consistent with the hypothesis that insufficient amounts of brain EPA and DHA may increase the susceptibility to impulsiveness and violence and therefore to violent suicide. Indeed, Dr. R. K. McNamara has found that DHA levels in postmortem samples of prefrontal cortex taken from suicide victims were lower than in control samples (McNamara et al. 2013).

In 1990, shortly before comparable studies on depression, a vast survey on the lifestyle in 265,000 Japanese followed for 17 years has shown that the daily consumption of one fish serving was associated with a 20% lower risk of suicide compared to subjects consuming less fish (Hirayama 1990). Later, a similar relationship was found in a population of 3000 individuals from Kuopio in the east-central Finland, where the subjects with the least suicidal thoughts consumed at least two fish servings per week (Tanskanen et al. 2001). In several countries concerned by their high suicide rate, these large epidemiological studies have formed the starting point of a lot of research.

The Japanese population, known for its high suicide rate, but also for its important fish consumption, has been the subject of a large epidemiological study that followed for 5 years 47,351 men and 54,156 women aged from 40 to 69 years (Poudel-Tandukar et al. 2011). A relatively high risk of death by suicide was observed in women only consuming very few fish. Therefore, the elevated national average of the ω-3 fatty acid intake is likely the cause of these unconvincing results, but the authors also refer to the diversity of social, nutritional, and psychiatric situations in the population. Equivocal results have also been reported in the United States after an extensive survey of the dietary habits and the number of suicides for 4 years in more than 205,000 subjects (Tsai et al. 2014).

At the European level, it is very difficult to establish any relationship between ω-3 fatty acids and suicide rates, with the official statistics and the surveys concerning nutrition in different countries being incomplete. In contrast, it is possible to take into account the overall amounts of the dietary polyunsaturated fatty acids (PUFAs) in 14 countries (Eilander et al. 2015). These values, published between 2004 and 2012, correspond mainly to the ingested linoleic acid (ω-6), the ω-3 fatty acids being 10–30 times lower. The official suicide statistics in Europe are accessible from the European Health for All database established for WHO (http://data.euro.who.int/hfadb/). From all the reliable and recent data on 14 countries, it is possible to show a positive correlation between the levels of PUFAs (as percentage of the total energy intake) and suicide rates (Figure 6.3). This

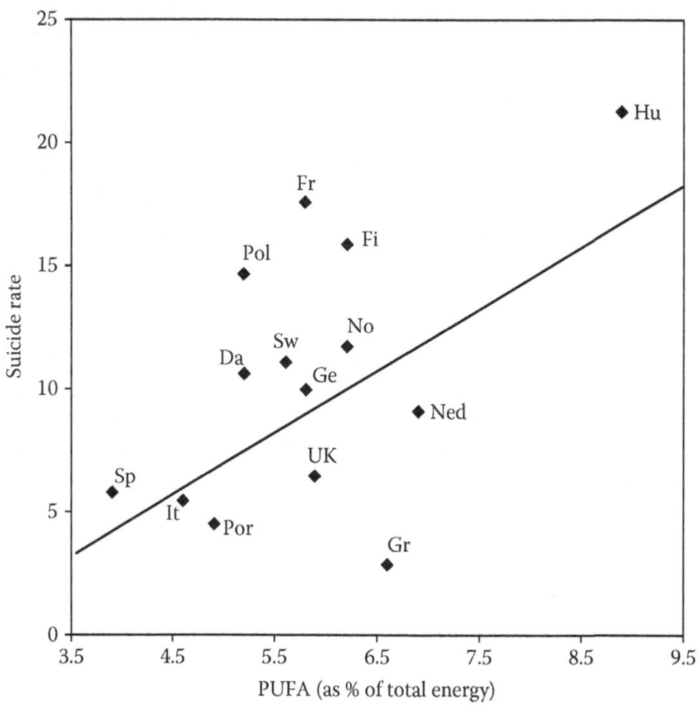

Figure 6.3 Relationship between suicide rates and the importance of dietary ω-6 fatty acids for people from 14 European countries ($R = 0.55$, $\alpha < 0.05$). Suicide rates (WHO source) are expressed per 100,000 inhabitants and the proportion of PUFA in percentage of the total energy intake (Eilander et al. 2015). Ge, Germany; De, Denmark; Sp, Spain; Fi, Finland; Fr, France; Gr, Greece; Ne, The Netherlands; Hu, Hungary; It, Italy; No, Norway; Pol, Poland; Por, Portugal; Sw, Sweden; UK: United Kingdom.

relationship highlights again the influence of the imbalance between the intake of ω-6 and ω-3 fatty acids in Western populations. All experts agree that the imbalance results from a too high intake of vegetable oils containing only ω-6 fatty acids (sunflower, peanut, or grape seed oil) at the expense of oils rich in ω-3 fatty acids (canola, walnut, flax oil, fish lipids). It would be important across Europe to collect reliable data on the amounts of the consumed ω-3 fatty acids, or better the dietary ω-6/ω-3 fatty acid ratio, to clarify their real influence on the suicidal behavior of populations. The situation in Greece (low suicide rates and high dietary PUFAs proportion) could thus be explained by the relative importance of the ω-3 fatty acids among all the dietary fatty acids.

These large surveys have their limits as a low accuracy being not necessarily compensated by the size of the cohorts studied. It is further increasingly clear that in a large-scale study, the origin of the ω-3 fatty acids and their consumed amounts are very difficult to appreciate. So, as in the other described studies, it is evidently more reliable to record internal biological parameters, reflecting more faithfully the typical diet of the investigated subjects.

To overcome these drawbacks, a Chinese study compared the fatty acid composition of erythrocytes in 100 subjects hospitalized for suicide attempts and in the same number of control subjects hospitalized after an accident. The authors observed a highly significant association between a low cellular EPA concentration and a high risk of suicide (Huan et al. 2004). Investigators at the Columbia University, New York, followed for 2 years 33 depressed subjects with high suicide risk (Sublette et al. 2006). After the observation of seven attempts of suicide, they found that the suicide risk was closely linked to low DHA levels in blood (Table 6.1) or to high levels of the ω-6/ω-3 fatty acid ratio. The authors did not hesitate to proclaim that blood DHA concentrations could predict future risks of suicide in depressed patients.

Table 6.1 Time course change of survival probability after suicide attempts based on the blood DHA level

Time to first attempts (days)	Low DHA (n = 17)	High DHA (n = 16)
0	100	100
200	86	100
400	86	100
600	71.3	100
800	51.5	93.4

Source: Sublette, M.E., et al. *Am. J. Psychiatry*, 163, 1100–1102, 2006.

A study done in the Galway National University, Ireland, compared 40 patients urgently admitted for suicide attempt with 40 control subjects having no psychiatric history (Garland et al. 2007). The authors confirmed that the diseased subjects differed significantly from controls, with the former having 47% lower EPA and 27.4% lower DHA and with cholesterol being 14% lower in plasma lipids.

Recently, a study of 27 patients with bipolar disorder showed that the 17 subjects with suicidal history had blood EPA levels slightly lower and arachidonic acid (20:4 ω-6) levels significantly lower than in the 10 subjects having no disorders (Evans et al. 2012). The analysis of fatty acid profiles enabled to conclude that the balance between the fatty acid precursor linolenic acid and its metabolites EPA and DHA was closely associated with the parameters measuring personality and suicidal intentions. Recently, a biochemical study of several metabolites (metabolomics study) demonstrated in bipolar subjects a similar relationship between fatty acid metabolism and psychiatric disorders (Evans et al. 2014). The precise analysis of these phenomena should make possible to consider later the opportunities for nutritional or pharmacological interventions in subjects at risk.

In response to a significantly increased number of suicides in soldiers, the US authorities undertook a major study to prevent the risk of suicide during or after operations. The blood ω-3 fatty acid composition in 800 soldiers whose death was by suicide was compared to that measured in 800 military in good mental health (Lewis et al. 2011). The investigators discovered that soldiers in front from 2002 to 2008 with the lowest DHA levels had a risk of suicide increased by 62% compared to soldiers with the highest fatty acid levels. The significance of these findings has motivated the US Army Medical Service to launch a major research program on this subject. The officials have already recommended a change in the military food rations in approaching closer to the essential fatty acid composition of the Mediterranean diet (Hibbeln and Gow 2014). The interest of military members in that prevention may surely grow as a group of doctors from the US Army has decided as part of a search for a "nutritional armor" to raise in an ethical way the ω-3 fatty acid status of military personnel.

If these results are confirmed by further studies, they could have implications for neurobiology of suicide and the ways psychiatrists will implement to reduce the risk of suicide, or at least the risk of a renewal of a previous attempt.

So far, no studies of prevention or treatment by using supplementation with ω-3 fatty acids have explored its possible effects on suicide frequency. In all investigations, including those referred to as foresight, there are several possible confounding factors (social status, favorable lifestyle, higher living standards, income) that may be associated with fish consumption, these factors being now increasingly taken into account.

Only one clinical study on the impact of ω-3 fatty acids on the resilience capacities of US veterans is listed on the NIH Clinical Trials website (http://clinicaltrials.gov).

6.2.2.2 *Vitamin D*

As mentioned at the beginning of section 6.2.2, the risk of a suicidal behavior increases dramatically in individuals with a mood disorder, even minor, combined with bipolar disorder or schizophrenia. In adolescents and young adults, it has been determined that nearly 80% of individuals who made at least one suicide attempt were suffering from one or several psychiatric disorders (Wunderlich et al. 1997).

Many studies have shown a close association between vitamin D and aggressive behavior (Section 6.2.1.2), but also several psychiatric disorders such as depression (Section 6.1.1.2), schizophrenia (Section 6.1.3.2), ADHD (Section 6.1.4.2), or autism (Section 6.1.5.2). It seems thus logical to add the suicidal behavior to this list (Tariq et al. 2011).

Until now, little work had been devoted to the possible correlations between the vitamin D status and the suicide risk. Therefore, a large survey was conducted in the United States by comparing blood vitamin D levels in soldiers who committed a fatal suicide with those of healthy military (Umhau et al. 2013). It seemed first that soldiers were very frequently vitamin D deficient, a probable consequence of their lifestyle due to their clothing; and second, the most deficient (<15 ng/mL) subjects had a high suicide risk. Because no relationship has been found between suicide and depression, the authors suggested that the aggressiveness linked to vitamin D deficiency in that population of soldiers could very naturally lead to the fatal act in connection with a reduction of the brain serotonergic activity (Lindqvist et al. 2010).

In Sweden, the presence of very low vitamin D levels (approximately 18 ng/mL) was observed among individuals who have attempted suicide but not followed by death, with that average vitamin level being 25% lower than in nonsuicidal depressed and 28% lower than in healthy controls (Grudet et al. 2014). In addition, 58% of suicidal subjects had a deficiency because their plasma vitamin D level was lower than the accepted threshold in most countries. The interesting nature of that work was to measure simultaneously several inflammation indicators. Thus, the authors have found that the vitamin D deficiency of suicidal subjects was accompanied by increased levels of proinflammatory cytokines (interleukin-6 and –1β). The known links between inflammatory processes and vitamin D (Zhang et al. 2012) will certainly boost the research, already very active in the field of suicidal behavior, at least in patients at risk suffering also from depression.

Given the metabolic link between vitamin D biosynthesis and sunshine, any seasonal variation of suicides offers a major way for the study of involved causes, allowing to consider preventative measures.

While the causes of suicide and the methods used to achieve it are very diverse, an annual frequency cycle has been observed with a peak at the end of spring and early summer and that in many European, American, or Asian countries (Christodoulou et al. 2012). When the precise circumstances of the suicide are known, the investigators have noticed that the cycle was more pronounced if the death was caused by violent methods (e.g., hanging, firearm).

Despite the apparent reality of the seasonal variations in suicides, there are conflicting studies raising the question of a possible interference of sex, age, suicide method, or other elusive sociological or biological parameters. Some epidemiologists have even evoked an antidepressant action of the sun in May–June strengthening the motivation to perform an act previously buried in the subconscious.

A broader survey of lifestyle along with vitamin D and serotonin assays could improve our knowledge on the physiological mechanisms underlying these biological cycles.

As with many psychiatric disorders mentioned above, a synchrony was observed between birth season and suicide frequency. After the first studies on small groups (Lester et al. 1970), the phenomenon was particularly analyzed in England, Wales, and Scotland on a large cohort of about 52,000 suicides listed for 21 years (Salib and Cortina-Borja 2010). In these countries, a 15% increase of the suicide risk for men and 27% for women was recorded for people born in April–May compared to those born in October–November. A recent study in Finland, including 1902 cases of suicide in patients with schizophrenia or other psychotic disorders reached similar conclusions (Karhumaa et al. 2013). This apparent predisposition to suicide in schizophrenic people born in the spring raises the problem of a possible role of vitamin D deficiency, likely due to a limited sun exposure during the second and third trimesters.

A causal link between this maternal deficiency and the onset of a suicidal behavior in adults has been considered by various authors, but it is still hypothetical as for other psychiatric or neurological disorders. Larger studies, and in several countries located in various latitudes, are needed before giving credit to that influence. Thus, prevention campaigns could help reduce suicidal risks, considering the potentiality of a long-term preventive treatment with vitamin D to decrease mortality after several psychiatric disorders.

6.2.2.3 Cholesterol

For the first time, in 1966, information was published on possible relationships between cholesterol and suicide (Hoch-Ligeti 1966). An exploration of the adrenal glands of subjects that died by suicide has revealed that their cholesterol content was about 60% higher than in subjects who died in accidents and not very different from that measured in subjects with hypertension. By contrast, along abundant research attempting to elucidate the relationships between cholesterol and cardiovascular disease, clinicians have noted an increased suicidal risk in groups of individuals having low blood cholesterol concentrations.

These first results were verified 24 years later in a large study involving 25,000 men followed for nearly 5 years, with half having a dietary plan or a treatment (without statin) to lower cholesterolemia (Muldoon et al. 1990). The authors observed a significant death increase not related to illness (accident, suicide, or violence) in subjects subjected to cholesterol lowering, but a lower mortality in subjects with coronary heart disease.

As for aggressive behavior (Section 6.2.1.3), the concern about these conclusions was noticed by the medical world that fostered several research studies to decide whether the new statin treatments fighting against hypercholesterolemia could induce a suicidal behavior, despite their interest and efficiency for cardiovascular diseases.

The first investigation in this field was already done in Sweden with more than 54,000 subjects followed for 20 years (Lindberg et al. 1992). Authors had verified that the lower the cholesterol level, the higher the mortality by suicide, but in men only. Thus, the relative suicide risk in the group of subjects having an average of 2 g/L of total cholesterol was nearly four times the risk measured in the group with an average of 2.9 g/L. The same year, an American study confirmed approximately these results after monitoring 351,000 men for 12 years (Neaton et al. 1992). Apart from a few exceptions, all subsequent studies have reached similar conclusions (Colin 2003). Very recently, a broad analysis of 65 studies published since 1994 and involving observations on more than 510,000 participants showed that those with the lowest cholesterol levels were significantly associated with a nearly two times suicide risk compared to those with high blood cholesterol levels (Wu et al. 2015).

Alarmed by these outcomes, some clinicians asked early on for a moratorium on the use of statins, the most widely used drugs to lower cholesterolemia (Smith and Pekkanen 1992). Fortunately, Prof. Dr. F. Muldoon, University of Pittsburgh, Pennsylvania, published in 2001 a review of 19 studies and concluded that there was currently no evidence of an effect of statins on mortality by suicide, whereas the other treatments (diet, various medications) should be avoided if possible (Muldoon et al. 2001).

All these results have been disputed, especially in the case of experiments for therapeutic purposes where the proposed diets contained lipids with different fatty acid compositions. In 1995, an analysis of the literature by J. R. Hibbeln had already highlighted the interference of the ω-6/ω-3 fatty acid ratio in the experiments (Hibbeln and Salem 1995). In fact, a lowering of cholesterolemia by a dietary change often leads to replace saturated fats by excessive amounts of vegetable oils, rich in ω-6 fatty acids. These changes clearly contribute to lower the relative contribution of ω-3 fatty acids, an alteration known to increase the depression risk (Section 6.1.1.1), and therefore suicide (Section 6.2.2.1).

This example illustrates the complex interconnections between dietary lipids, physiology, and the behavior of individuals, beneficial modifications for a system being able to hamper one or several others.

The direct relationships between cholesterol and suicide attempts were clinically explored several times. A study of people hospitalized with bipolar disorder has shown that those who attempted suicide had a 20% lower cholesterol level compared to those who have made no attempt (Vuksan-Cusa et al. 2009). Interestingly, when hypocholesterolemia is of familial origin, due to a mutation of a specific lipoprotein, the affected subjects have a very high risk of violent death by suicide (Edgar et al. 2007).

A study done in Prague, Czech Republic, has clarified that cholesterol levels measured in women hospitalized after suicide attempt were significant lower compared to controls when violent means only had been used (stabbing, firearm, hanging, drowning, falling). Subjects using nonviolent means (drugs) were comparable to controls (Vevera et al. 2003). Despite the absence of dietary data, the present results clearly show that subjects attempting suicide are not a homogeneous group. The means used for their suicide attempt enable the distinction of at least two types of subjects, likely related to a change in brain serotonergic activity (Mann 1989).

Another discovery highlighted the complexity of the family behavior combining violence and suicide, parents and children. Indeed, it has been shown that among a group of people who attempted suicide by violent means, those having a cholesterol level below the median of the group had been more frequently exposed to violence in their childhood (6–14 years) than individuals with a cholesterol level above the median (Asellus et al. 2014). Thus, it seems that the circulating cholesterol level may change the "cycle of violence" often described in violent adults who have experienced violent periods in their childhood. It is likely that if that cycle really exists, it seems more significant in individuals with low cholesterol levels. Unfortunately, there is no record in children to assess the influence of their cholesterolemia on their future behavior; conversely, it is the same for a possible effect of violent behaviors in childhood on cholesterol metabolism.

Although the links between cholesterol and suicidal behavior are now accepted by the scientific community, the mechanisms involved remain poorly defined. Most hypotheses suggest, however, the intervention of a reduced serotonergic activity, likely a consequence of a lack of cholesterol at the level of synapses. The increased impulsivity could find its origin from other mechanisms, such as a regulation by steroids or by some neurotrophic factors (Cantarelli et al. 2014).

To date, only two studies on the links between cholesterol and suicidal behavior are listed in the US Government site for clinic trails (https://clinicaltrials.gov/).

References

Adams, P.B., Lawson, S., Sanigorski, A., et al. 1996. Arachidonic acid to eicosapentaenoic acid ratio in blood correlates positively with clinical symptoms of depression. *Lipids* 31:157–61.

Akter, K., Gallo, D.A., Martin, S.A., et al. 2012. A review of the possible role of the essential fatty acids and fish oils in the aetiology, prevention or pharmacotherapy of schizophrenia. *J. Clin. Pharm. Ther.* 37:132–9.

Amato, R., Pinelli, M., Monticelli, A., et al. 2010. Schizophrenia and vitamin D related genes could have been subject to latitude-driven adaptation. *BMC Evol. Biol.* 10:351.

Amminger, G.P., Berger, G.E., Schäfer, M.R., et al. 2007. Omega-3 fatty acids supplementation in children with autism: A double-blind randomized, placebo-controlled pilot study. *Biol. Psychiatry* 61:551–3.

Amminger, G.P., Schäfer, M.R., Papageorgiou, K., et al. 2010. Long-chain omega-3 fatty acids for indicated prevention of psychotic disorders: A randomized, placebo-controlled trial. *Arch. Gen. Psychiatry* 67:146–54.

Appleton, K.M., Sallis, H.M., Perry, R., et al. 2016. ω-3 Fatty acids for major depressive disorder in adults: An abridged Cochrane review. *BMJ Open* 6:e010172.

Arvindakshan, M., Ghate, M., Ranjekar, P.K., et al. 2003. Supplementation with a combination of omega-3 fatty acids and antioxidants (vitamins E and C) improves the outcome of schizophrenia. *Schizophr. Res.* 62:195–204.

Asberg, M., Träskman, L., Thorén, P. 1976. 5-HIAA in the cerebrospinal fluid. A biochemical suicide predictor? *Arch. Gen. Psychiatry* 33:1193–7.

Asellus, P., Nordström, P., Nordström, A.L., et al. 2014. Cholesterol and the "Cycle of Violence" in attempted suicide. *Psychiatry Res.* 215:646–50.

Ataie-Jafari, A., Qorbani, M., Heshmat, R., et al. 2015. The association of vitamin D deficiency with psychiatric distress and violence behaviors in Iranian adolescents: The CASPIAN-III study. *J. Diabetes Metab. Disord.* 14:62.

Baek, D., Park, Y. 2013. Association between erythrocyte n-3 polyunsaturated fatty acids and biomarkers of inflammation and oxidative stress in patients with and without depression. *Prostaglandins Leukot. Essent. Fatty Acids* 89:291–6.

Baird, G., Simonoff, E., Pickles, A., et al. 2006. Prevalence of disorders of the autism spectrum in a population cohort of children in South Thames: The Special Needs and Autism Project (SNAP). *Lancet* 368(9531):210–15.

Barragán, E., Breuer, D., Döpfner, M. 2014. Efficacy and safety of omega-3/6 fatty acids, methylphenidate, and a combined treatment in children with ADHD. *J. Atten. Disord.* 21:433–41.

Beck, A.T., Kovacs, M., Weissman, A. 1979. Assessment of suicidal intention: The scale of suicide ideation. *J. Consult. Clin. Psychol.* 47:343–52.

Beier, A.M., Lauritzen, L., Galfalvy, H.C., et al. 2014. Low plasma eicosapentaenoic acid levels are associated with elevated trait aggression and impulsivity in major depressive disorder with a history of comorbid substance use disorder. *J. Psychiatr. Res.* 57:133–40.

Bell, J.G., Miller, D., MacDonald, D.J., et al. 2010. The fatty acid compositions of erythrocyte and plasma polar lipids in children with autism, developmental delay or typically developing controls and the effect of fish oil intake. *Br. J. Nutr.* 103:1160–7.

Bell, J.G., Sargent, J.R., Tocher, D.R., et al. 2000. Red blood cell fatty acid compositions in a patient with autistic spectrum disorder: A characteristic abnormality in neurodevelopmental disorders? *Prostaglandins Leukot. Essent. Fatty Acids* 63:21–5.

Belvederi Muri, M., Respino, M., Masotti, M. et al. 2013.Vitamin D and psychosis: Mini metaanalysis. *Schizophr. Res.* 150:235–9.

Belzeaux, R., Boyer, L., Ibrahim, E.C., et al. 2015. Mood disorders are associated with a more severe hypovitaminosis D than schizophrenia. *Psychiatry Res.* 229:613–16.

Bener, A., Kamal, M. 2014. Predict attention deficit hyperactivity disorder? Evidence-based medicine. *Glob. J. Health Sci.* 6:47–57.

Bent, S., Bertoglio, K., Ashwood, P., et al. 2011. A pilot randomized controlled trial of omega-3 fatty acids for autism spectrum disorder. *J. Autism Dev. Disord.* 41:545–54.

Bentsen, H., Solberg, D.K., Refsum, H., et al. 2012. Clinical and biochemical validation of two endophenotypes of schizophrenia defined by levels of polyunsaturated fatty acids in red blood cells. *Prostaglandins Leukot. Essent. Fatty Acids* 87:35–41.

Bertone-Johnson, E.R. 2009. Vitamin D and the occurrence of depression: Causal association or circumstantial evidence? *Nutr. Rev.* 67:481–92.

Beydoun, M.A., Fanelli Kuczmarski, M.T., Beydoun, H.A., et al. 2013. ω-3 Fatty acid intakes are inversely related to elevated depressive symptoms among United States women. *J. Nutr.* 143:1743–52.

Bigornia, S.J., Harris, W.S., Falcón, L.M., et al. 2016. The omega-3 index is inversely associated with depressive symptoms among individuals with elevated oxidative stress biomarkers. *J. Nutr.* 146:758–66.

Bilici, M., Efe, H., Köroğlu, M.A., et al. 2001. Antioxidative enzyme activities and lipid peroxidation in major depression: Alterations by antidepressant treatments. *J. Affect. Dis.* 64:43–51.

Boland, M.R., Shahn, Z., Madigan, D., et al. 2015. Birth month affects lifetime disease risk: A phenome-wide method. *J. Am. Med. Inform. Assoc.* 22:1042–53.

Bos, D.J., Oranje, B., Veerhoek, E.S., et al. 2015. Reduced symptoms of inattention after dietary omega-3 fatty acid supplementation in boys with and without attention deficit/hyperactivity disorder. *Neuropsychopharmacology* 40:2298–306.

Brown, G.L., Goodwin, F., Ballenger, J., et al. 1979. Aggression in humans correlates with cerebrospinal fluid amine metabolites. *Psychiatry Res.* 1:131–9.

Burgess, J.R., Stevens, L., Zhang, W., et al. 2000. Long-chain polyunsaturated fatty acids in children with attention-deficit hyperactivity disorder. *Am. J. Clin. Nutr.* 71:327S–30S.

Buydens-Branchey, L., Branchey, M., McMakin, D.L., et al. 2003a. Polyunsaturated fatty acid status and aggression in cocaine addicts. *Drug Alcohol Depend.* 71:319–23.

Buydens-Branchey, L., Branchey, M., McMakin, D.L., et al. 2003b. Polyunsaturated fatty acid status and relapse vulnerability in cocaine addicts. *Psychiatry Res.* 120:29–35.

Cannell, J.J. 2008. Autism and vitamin D. *Med. Hypothesis* 70:750–9.

Cantarelli, M.G., Tramontina, A.C., Leite, M.C., et al. 2014. Potential neurochemical links between cholesterol and suicidal behavior. *Psychiatry Res.* 220:745–51.

Chevreul, K., Prigent, A., Bourmaud, A., et al. 2013. The cost of mental disorders in France. *Eur. Neuropsychopharmacol.* 23:879–86.

Christensen, O., Christensen, E. 1988. Fat consumption and schizophrenia. *Acta Psychiatr. Scand.* 78:587–91.

Christodoulou, C., Douzenis, A., Papadopoulos, F.C., et al. 2012. Suicide and seasonality. *Acta Psychiatr. Scand.* 125:127–46.

Chung, H., Choi, A., Cho, I.H., et al. 2008. Changes in fatty acids and volatile components in mackerel by broiling. *Eur. J. Lipid Technol.* 113:1481–90.

Clayton, E.H., Hanstock, T.L., Hirneth, S.J., et al. 2008. Long-chain omega-3 polyunsaturated fatty acids in the blood of children and adolescents with juvenile bipolar disorder. *Lipids* 43:1031–8.

Clelland, J.D., Read, L.L., Drouet, V., et al. 2014. Vitamin D insufficiency and schizophrenia risk: Evaluation of hyperprolinemia as a mediator of association. *Schizophr. Res.* 156:15–22.

Colin, A. 2003. Lipids, depression and suicide. *Encéphale* 29:49–58.

Colquhoun, I., Bunday, S. 1981. A lack of essential fatty acids as a possible cause of hyperactivity in children. *Med. Hypoth.* 7:673–9.

Corrigan, F., Gray, R., Strathdee, A., et al. 1994. Fatty acid analysis of blood from violent offenders. *J. Forens. Psychiatry Psychol.* 5:83–92.

Cremniter, D., Jamain, S., Kollenbach, K., et al. 1999. CSF 5-HIAA levels are lower in impulsive as compared to nonimpulsive violent suicide attempters and control subjects. *Biol. Psychiatry* 45:1572–9.

Da Rocha, C.M., Kac, G. 2012. High dietary ratio of omega-6 to omega-3 polyunsaturated acids during pregnancy and prevalence of post-partum depression. *Matern. Child Nutr.* 8:36–48.

Dealberto, M.J. 2011. Prevalence of autism according to maternal immigrant status and ethnic origin. *Acta Psychiatr. Scand.* 123:339–48.

Dean, A.J., Bor, W., Adam, K., et al. 2014. A randomized, controlled, crossover trial of fish oil treatment for impulsive aggression in children and adolescents with disruptive behavior disorders. *J. Child Adolesc. Psychopharmacol.* 24:140–8.

Derogatis, L.R., Lipman, R.S., Covi, L. 1973. SCL-90: An outpatient psychiatric rating scale—preliminary report. *Psychopharmacol. Bull.* 9:13–28.

De Vriese, S.R., Christophe, A.B., Maes, M. 2004. In humans, the seasonal variation in polyunsaturated fatty acids is related to the seasonal variation in violent suicide and serotonergic markers of violent suicide. *Prostaglandins Leukot. Essent. Fatty Acids* 71:13–18.

Disanto, G., Morahan, J.M., Lacey, M.V., et al. 2012. Seasonal distribution of psychiatric births in England. *PLoS One* 7:e34866.

Edgar, P.F., Hooper, A.J., Poa, N.R., et al. 2007. Violent behavior associated with hypocholesterolemia due to a novel APOB gene mutation. *Mol. Psychiatry* 12:258–63.

Eilander, A., Rajwinder, K., Harika, K., et al. 2015. Intake and sources of dietary fatty acids in Europe: Are current population intakes of fats aligned with dietary recommendations? *Eur. J. Lipid Sci. Technol.* 117:1370–7.

Evans, S.J., Prossin, A.R., Harrington, G.J., et al. 2012. Fats and factors: Lipid profiles associate with personality factors and suicidal history in bipolar subjects. *PLoS One* 7:e29297.

Evans, S.J., Ringrose, R.N., Harrington, G.J., et al. 2014. Dietary intake and plasma metabolomic analysis of polyunsaturated fatty acids in bipolar subjects reveal dysregulation of linoleic acid metabolism. *J. Psychiatr. Res.* 57:58–64.

Féart, C., Peuchant, E., Letenneur, L., et al. 2008. Plasma eicosapentaenoic acid is inversely associated with severity of depressive symptomatology in the elderly: Data from the Bordeaux sample of the Three-City Study. *Am. J. Clin. Nutr.* 87:1156–62.

Fernell, E., Barnevik-Olsson, M., Bågenholm, G., et al. 2010. Serum levels of 25-hydroxyvitamin D in mothers of Swedish and of Somali origin who have children with and without autism. *Acta Paediatr.* 99:743–7.

Fernell, E., Bejerot, S., Westerlund, J., et al. 2015. Autism spectrum disorder and low vitamin D at birth: A sibling control study. *Mol. Autism* 6:3.

Frangou, S., Lewis, M., McCrone, P. 2006. Efficacy of ethyl-eicosapentaenoic acid in bipolar depression: Randomised double-blind placebocontrolled study. *Br. J. Psychiatry* 188:46–50.

Frensham, L.J., Bryan, J., Parletta, N. 2012. Influences of micronutrient and omega-3 fatty acid supplementation on cognition, learning, and behavior: Methodological considerations and implications for children and adolescents in developed societies. *Nutr. Rev.* 70:594–610.

Ganji, V., Milone, C., Cody, M.M., et al. 2010. Serum vitamin D concentrations are related to depression in young adult US population: The third national health and nutrition examination survey. *Int. Arch. Med.* 3:29.

Garcion, E., Wion-Barbot, N., Montero-Menei, C.N., et al. 2002. New clues about vitamin D functions in the nervous system. *Trends Endocrinol. Metab.* 13:100–5.

Garland, M.R., Hallahan, B., McNamara, M., et al. 2007. Lipids and essential fatty acids in patients presenting with self-harm. *Br. J. Psychiatry* 190:112–17.

Gesch, C.B., Hammond, S.M., Hampson, S.E., et al. 2002. Influence of supplementary vitamins, minerals and essential fatty acids on the antisocial behaviour of young adult prisoners. Randomised, placebo-controlled trial. *Br. J. Psychiatry* 181:22–8.

Gillberg, C., Wing, L. 1999. Autism: Not an extremely rare disorder. *Acta Psychiatr. Scand.* 99:399–406.

Gloth, F.M., Alam, W., Hollis, B. 1999. Vitamin D vs broad spectrum phototherapy in the treatment of seasonal affective disorder. *J. Nutr. Health Aging* 3:5–7.

Goksugur, S.B., Tufan, A.E., Semiz, M., et al. 2014. Vitamin D status in children with attention-deficit-hyperactivity disorder. *Pediatr. Int.* 56:515–19.

Golomb, B.A., Stattin, H., Mednick, S. 2000. Low cholesterol and violent crime. *J. Psychiatr. Res.* 34:301–9.

Gow, R.V., Sumich, A., Vallee-Tourangeau, F., et al. 2013. Omega-3 fatty acids are related to abnormal emotion processing in adolescent boys with attentiondeficit hyperactivity disorder. *Prostaglandins Leukot. Essent. Fatty Acids* 88:419–29.

Grant, W.B., Cannell, J.J. 2013. Autism prevalence in the United States with respect to solar UV-B doses: An ecological study. *Dermato-Endocrinology* 5:159–64.

Grant, W.B., Soles, C.M. 2009. Epidemiologic evidence supporting the role of maternal vitamin D deficiency as a risk factor for the development of infantile autism. *Dermato-Endocrinology* 1:223–8.

Grosso, G., Pajak, A., Marventano, S., et al. 2014. Role of omega-3 fatty acids in the treatment of depressive disorders: A comprehensive meta-analysis of randomized clinical trials. *PLoS One* 9:e96905.

Grudet, C., Malm, J., Westrin, A., et al. 2014. Suicidal patients are deficient in vitamin D, associated with a pro-inflammatory status in the blood. *Psychoneuroendocrinology* 50:210–19.

Hamazaki, T., Hamazaki, K. 2008. Fish oils and aggression or hostility. *Prog. Lipid Res.* 47:221–32.

Hamazaki, T., Sawazaki, S., Itomura, M., et al. 1996. The effect of docosahexaenoic acid on aggression in young adults. A placebo-controlled double-blind study. *J. Clin. Invest.* 97:1129–33.

Hamazaki, T., Sawazaki, S., Itomura, M., et al. 2001. Effect of docosahexaenoic acid on hostility. *World Rev. Nutr. Diet.* 88:47–52.

Hedelin, M., Löf, M., Olsson, M., et al. 2010. Dietary intake of fish, omega-3, omega-6 polyunsaturated fatty acids and vitamin D and the prevalence of psychotic-like symptoms in a cohort of 33,000 women from the general population. *BMC Psychiatry* 10:38.

Hibbeln, J.R. 1998. Fish consumption and major depression. *Lancet* 351:1213.

Hibbeln, J.R. 2001. Seafood consumption and homicide mortality. A cross-national ecological analysis. *World Rev. Nutr. Diet* 88:41–6.

Hibbeln, J.R. 2007. From homicide to happiness—a commentary on omega-3 fatty acids in human society. *Nutr. Health* 19:9–19.

Hibbeln, J.R., Nieminen, L.R., Lands, W.E. 2004a. Increasing homicide rates and linoleic acid consumption among five Western countries, 1961–2000. *Lipids* 39:1207–13.

Hibbeln, J.R., Bissette, G., Umhau, J.C., et al. 2004b. Omega-3 status and cerebrospinal fluid corticotrophin releasing hormone in perpetrators of domestic violence. *Biol. Psychiatry* 56:895–7.

Hibbeln, J.R., Gow, R.V. 2014. The potential for military diets to reduce depression, suicide, and impulsive aggression: A review of current evidence for omega-3 and omega-6 fatty acids. *Mil. Med.* 179:117–28.

Hibbeln, J.R., Linnoila, M., Umhau, J.C., et al. 1998. Essential fatty acids predict metabolites of serotonin and dopamine in cerebrospinal fluid among healthy control subjects, and early-and late-onset alcoholics. *Biol. Psychiatry* 44:235–42.

Hibbeln, J.R., Salem, N. 1995. Dietary polyunsaturated fatty acids and depression: When cholesterol does not satisfy. *Am. J. Clin. Nutr.* 62:1–9.

Hillbrand, M., Spitz, R.T., Foster, H.G. 1995. Serum cholesterol and aggression in hospitalized male forensic patients. *J. Behav. Med.* 18:33–43.

Hirayama, T. 1990. *Life-style and mortality: A large census-based cohort study in Japan.* Basel: Karger.

Hoch-Ligeti, C. 1966. Adrenal cholesterol concentration in cases of suicide. *Br. J. Exp. Pathol.* 47:594–8.

Hoffmire, C.A., Block, R.C., Thevenet-Morrison, K., et al. 2012. Associations between omega-3 polyunsaturated fatty acids from fish consumption and severity of depressive symptoms: An analysis of the 2005–2008 National Health and Nutrition Examination Survey. *Prostaglandins Leukot. Essent. Fatty Acids* 86:155–60.
Holman, R.T., Johnson, S.B., Hatch, T.F. 1982. A case of human linolenic acid deficiency involving neurological abnormalities. *Am. J. Clin. Nutr.* 35:617–23.
Horikawa, C., Otsuka, R., Kato, Y., et al. 2016. Cross-sectional association between serum concentrations of n-3 long-chain PUFA and depressive symptoms: Results in Japanese community dwellers. *Br. J. Nutr.* 115:672–80.
Horrobin, D.F. 1977. Schizophrenia as a prostaglandin deficiency disease. *Lancet* 1 (8018):936–7.
Horrobin, D.O. 1998. Schizophrenia: The illness that made us human. *Med. Hypotheses* 50:269–88.
Horrobin, D.F., Manku, M.S., Hillman, H., et al. 1991. Fatty acid levels in the brains of schizophrenics and normal controls. *Biol. Psychiatry* 30:795–805.
Horrobin, D.F., Manku, M.S., Morse-Fisher, N., et al. 1989. Essential fatty acids in plasma phospholipids in schizophrenics. *Biol. Psychiatry* 25:562–8.
Horta, B.L. and Victora, C.G. 2013. Long-term effects of breastfeeding. A systematic review. WHO. http://apps.who.int/iris/bitstream/10665/79198/1/9789241505307_eng.pdf
Huan, M., Hamazaki, K., Sun, Y., et al. 2004. Suicide attempt and n-3 fatty acid levels in red blood cells: A case control study in China. *Biol. Psychiatry* 56:490–6.
Husemoen, L.L., Ebstrup, J.F., Mortensen, E.L., et al. 2016. Serum 25-hydroxyvitamin D and self-reported mental health status in adult Danes. *Eur. J. Clin. Nutr.* 70:78–84.
INPES 2017. http://inpes.santepubliquefrance.fr/70000/cp/07/cp071009.asp.
Iribarren, C., Markovitz, J.H., Jacobs, D.R., et al. 2004. Dietary intake of n-3, n-6 fatty acids and fish: Relationship with hostility in young adults—the CARDIA study. *Eur. J. Clin. Nutr.* 58:24–31.
Itomura, M., Hamazaki, K., Sawazaki, S., et al. 2005. The effect of fish oil on physical aggression in schoolchildren—a randomized, double-blind, placebo-controlled trial. *J. Nutr. Biochem.* 16:163–71.
Jääskeläinena, T., Knekt, P., Suvisaari, J., et al. 2015. Higher serum 25-hydroxyvitamin D concentrations are related to a reduced risk of depression. *Br. J. Nutr.* 113:1418–26.
Jacobs, E.T., Mullany, C.J. 2015. Vitamin D deficiency and inadequacy in a correctional population. *Nutrition* 31:659–63.
Jamilian, H., Solhi, H., Jamilian, M. 2014. Randomized, placebo-controlled clinical trial of omega-3 as supplemental treatment in schizophrenia. *Glob. J. Health Sci.* 6:103–8.
Jiang, P., Zhang, L.H., Cai, H.L., et al. 2014. Neurochemical effects of chronic administration of calcitriol in rats. *Nutrients* 6:6048–59.
Kaneko, I., Sabir, M.S., Dussik, C.M., et al. 2015. 1,25-Dihydroxyvitamin D regulates expression of the tryptophan hydroxylase 2 and leptin genes: Implication for behavioral influences of vitamin D. *FASEB J.* 29:4023–35.
Kaplan, J.R., Fontenot, M.B., Manuck, S.B., et al. 1996. Influence of dietary Lipids on agonistic and affiliative behavior in *Macaca fascicularis*. *Am. J. Primatol.* 38:333–47.

Karhumaa, T., Hakko, H., Nauha, R., et al. 2013. Season of birth in suicides: Excess of births during the summer among schizophrenic suicide victims. *Neuropsychobiology* 68:238–42.

Kerr, D.C., Zava, D.T., Piper, W.T., et al. 2015. Associations between vitamin D levels and depressive symptoms in healthy young adult women. *Psychiatry Res.* 227:46–51.

Kim, Y.S., Leventhal, B.L., Koh, Y.J., et al. 2011. Prevalence of autism spectrum disorders in a total population sample. *Am. J. Psychiatry* 168:904–12.

Kim, S.W., Schäfer, M.R., Klier, C.M., et al. 2014. Relationship between membrane fatty acids and cognitive symptoms and information processing in individuals at ultra-high risk for psychosis. *Schizophr. Res.* 158:39–44.

Kingston, D., Tough, S., Whitfield, H. 2012. Prenatal and postpartum maternal psychological distress and infant development: A systematic review. *Child Psychiatry Hum. Dev.* 43:683–714.

Kocovska, E., Fernell, E., Billstedt, E., et al. 2012. Vitamin D and autism: Clinical review. *Res. Dev. Disabil.* 33:1541–50.

Landfield, P.W., Cadwallader-Neal, L. 1998. Long-term treatment with calcitriol (1,25(OH)2 vit D3) retards a biomarker of hippocampal aging in rats. *Neurobiol. Aging* 19:469–77.

Laugharne, J.D., Mellor, J.E., Peet, M. 1996. Fatty acids and schizophrenia. *Lipids* 31:S163–5.

Le-Niculescu, H., Case, N.J., Hulvershorn, L., et al. 2011. Convergent functional genomic studies of ω-3 fatty acids in stress reactivity, bipolar disorder and alcoholism. *Transl. Psychiatry* 1:e4.

Leray, C. 2015. *Lipids. Nutrition and health.* Boca Raton, FL: CRC Press.

Lester, D., Reeve, C.L., Priebe, K. 1970. Completed suicide and month of birth. *Psychol. Rep.* 27:210.

Lewis, M.D., Hibbeln, J.R., Johnson, J.E., et al. 2011. Suicide deaths of active-duty US military and omega-3 fatty-acid status: A case-control comparison. *J. Clin. Psychiatry* 72:1585–90.

Li, F., Liu, X., Zhang, D. 2016. Fish consumption and risk of depression: A meta-analysis. *J. Epidemiol. Community Health* 70:299–304.

Lieb, J., Karmali, R., Horrobin, D. 1983. Elevated levels of prostaglandin E2 and thromboxane B2 in depression. *Prostaglandins Leukot. Med.* 10:361–7.

Lindberg, G., Råstam, L., Gullberg, B., et al. 1992. Low serum cholesterol concentration and short term mortality from injuries in men and women. *BMJ* 305:277–9.

Lindqvist, D., Janelidze, S., Erhardt, S., et al. 2010. CSF biomarkers in suicide attempters–a principal component analysis. *Acta Psychiatr. Scand.* 124:52–61.

Linnoila, M., Virkkunen, M., Scheinin, M., et al. 1983. Low cerebrospinal fluid 5-hydroxyindoleacetic acid concentration differentiates impulsive from nonimpulsive violent behavior. *Life Sci.* 33:2609–14.

Lippi, G., Bonelli, P., Buonocore, R., et al. 2015. Birth season and vitamin D concentration in adulthood. *Ann. Transl. Med.* 3:231.

London, E. 2000. The environment as an etiologic factor in autism: A new direction for research. *Environment. Health Perspect.* 108 Suppl 3:401–4.

Long, S.J., Benton, D. 2013. A doubleblind trial of the effect of docosahexaenoic acid and vitamin and mineral supplementation on aggression, impulsivity, and stress. *Hum. Psychopharmacol.* 28:238–47.

Lu, D.Y., Tsao, Y.Y., Leung, Y.M., et al. 2010. Docosahexaenoic acid suppresses neuroinflammatory responses and induces heme oxygenase-1 expression in BV-2 microglia: Implications of antidepressant effects for ω-3 fatty acids. *Neuropsychopharmacology* 35:2238–48.

Lyall, K., Munger, K.L., O'Reilly, É.J., et al. 2013. Maternal dietary fat intake in association with autism spectrum disorders. *Am. J. Epidemiol.* 178:209–20.

Maes, M., De Vos, N., Pioli, R., et al. 2000. Lower serum vitamin E concentrations in major depression. Another marker of lowered antioxidant defenses in that illness. *J. Affect. Dis.* 58:241–6.

Mann, J.J. 1989. Neurochemical studies of violent and nonviolent suicide. *Psychopharmacol. Bull.* 25:407–12.

Marí-Bauset, S., Llópis-González, A., Zazpe, I., et al. 2015. Fat intake in children with autism spectrum disorder in the Mediterranean region (Valencia, Spain). *Nutr. Neurosci.* 19(9):377–386.

Markhus, M.W., Skotheim, S., Graff, I.E., et al. 2013. Low omega-3 index in pregnancy is a possible biological risk factor for post-partum depression. *PLoS One* 8:e67617.

Mazahery, H., Camargo, C.A., Conlon, C., et al. 2016. Vitamin D and autism spectrum disorder: A literature review. *Nutrients* 8:236.

McGrath, J. 1999. Hypothesis: Is low prenatal vitamin D a risk-modifying factor for schizophrenia? *Schizophr. Res.* 40:173–7.

McGrath, J., Saari, K., Hakko, H., et al. 2004. Vitamin D supplementation during the first year of life and risk of schizophrenia: A Finnish birth cohort study. *Schizophr. Res.* 67:237–45.

McNamara, R.K., Hahn, C.G., Jandacek, R., et al. 2007a. Selective deficits in the omega-3 fatty acid docosahexaenoic acid in the postmortem orbitofrontal cortex of patients with major depressive disorder. *Biol. Psychiatry*. 62:17–24.

McNamara, R.K., Jandacek, R., Rider, T., et al. 2007b. Abnormalities in the fatty acid composition of the postmortem orbitofrontal cortex of schizophrenic patients: Gender differences and partial normalization with antipsychotic medications. *Schizophr. Res.* 91:37–50.

McNamara, R.K., Jandacek, R., Rider, T., et al. 2010. Selective deficits in erythrocyte docosahexaenoic acid composition in adult patients with bipolar disorder and major depressive disorder. *J. Affect. Disord.* 126:303–11.

McNamara, R.K., Jandacek, R., Tso, P., et al. 2013. Lower docosahexaenoic acid concentrations in the postmortem prefrontal cortex of adult depressed suicide victims compared with controls without cardiovascular disease. *J. Psychiatry Res.* 47:1187–91.

Meesters, C., Muris, P., Bosma, H., et al. 1996. Psychometric evaluation of the Dutch version of the Aggression Questionnaire. *Behav. Res. Ther.* 34:839–43.

Mellor, J.E., Laugharne, J.D., Peet, M. 1995. Schizophrenic symptoms and dietary intake of n-3 fatty acids. *Schizophr. Res.* 18:85–6.

Meyer, B.J., Byrne, M.K., Collier, C., et al. 2015. Baseline omega-3 index correlates with aggressive and attention deficit disorder behaviours in adult prisoners. *PLoS One* 10:e0120220.

Michalak, J., Zhang, X.C., Jacobi, F. 2012. Vegetarian diet and mental disorders: Results from a representative community survey. *Int. J. Behav. Nutr. Phys. Act.* 9:67.

Milte, C.M., Parletta, N., Buckley, J.D., et al. 2012. Eicosapentaenoic and docosahexaenoic acids, cognition, and behavior in children with attention-deficit/hyperactivity disorder: A randomized controlled trial. *Nutrition* 28:670–7.

Mizoue, T., Kochi, T., Akter, S., et al. 2015. Low serum 25-hydroxyvitamin D concentrations are associated with increased likelihood of having depressive symptoms among Japanese workers. *J. Nutr.* 145:541–6.

Moskovitz, R.A. 1978. Seasonality in schizophrenia. *Lancet* 311:664.

Mossin, M.H., Aaby, J.B., Dalgård, C., et al. 2016. Inverse associations between cord vitamin D and attention deficit hyperactivity disorder symptoms: A child cohort study. *Aust. N. Z. J. Psychiatry*. DOI: 10.1177/0004867416670013 (Epub ahead of print).

Mozaffari-Khosravi, H., Yassini-Ardakani, M., Karamati, M., et al. 2013. Eicosapentaenoic acid versus docosahexaenoic acid in mild-to-moderate depression: A randomized, double-blind, placebo-controlled trial. *Eur. Psychoneuropharmacol.* 23:636–44.

Mtabaji, J.P., Manku, M.S., Horrobin, D.F. 1977. Actions of the tricyclic antidepressant clomipramine on responses to pressor agents. Interactions with prostaglandin E2. *Prostaglandins* 14:273–81.

Muldoon, M.F., Manuck, S.B., Matthews, K.A. 1990. Lowering cholesterol concentrations and mortality: A quantitative review of primary prevention trials. *BMJ* 301:309–14.

Muldoon, M.F., Manuck, S.B., Mendelsohn, A.B., et al. 2001. Cholesterol reduction and nonillness mortality: Meta-analysis of randomised clinical trials. *BMJ* 322:11–15.

Nanri, A., Mizoue, T., Matsushita, Y., et al. 2009. Association between serum 25-hydroxyvitamin D and depressive symptoms in Japanese: Analysis by survey season. *Eur. J. Clin. Nutr.* 63:1444–7.

Neaton, J.D., Blackburn, H., Jacobs, D., et al. 1992. Serum cholesterol level and mortality findings for men screened in the multiple risk factor intervention. Trial. Multiple Risk Factor Intervention Trial Research Group. *Arch. Intern. Med.* 152:1490–500.

Nemets, B., Stahl, Z., Belmaker, R.H. 2002. Addition of omega-3 fatty acid to maintenance medication treatment for recurrent unipolar depressive disorder. *Am. J. Psychiatry* 159:477–9.

Noaghiul, S., Hibbeln, J.R. 2003. Cross-national comparisons of seafood consumption and rates of bipolar disorders. *Am. J. Psychiatry* 160:2222–7.

Ooi, Y.P., Weng, S.J., Jang, L.Y., et al. 2015. Omega-3 fatty acids in the management of autism spectrum disorders: Findings from an open-label pilot study in Singapore. *Eur. J. Clin. Nutr.* 69:969–71.

Opie, R.S., Itsiopoulos, C., Parletta, N., et al. 2015. Dietary recommendations for the prevention of depression. *Nutr. Neurosci.* Aug 28 (Epub ahead of print).

Ovesen, L., Andersen, R., Jakobsen, J. 2003. Geographical differences in vitamin D status, with particular reference to European countries. *Proc. Nutr. Soc.* 62:813–21.

Owen, A.J., Batterham, M.J., Probst, Y.C., et al. 2005. Low plasma vitamin E levels in major depression: Diet or disease? *Eur. J. Clin. Nutr.* 59:304–6.

Pariente, P., Smith, M. 1990. Detection of the anxio-depressive disorders in liaison psychiatry. Contribution of the General Health Questionnaire leur degré d'agressivité. *Encéphale* 16:459–64.

Park, Y., Park, Y.S., Kim, S.H., et al. 2015. Supplementation of n-3 polyunsaturated fatty acids for major depressive disorder: A randomized, double-blind, 12-week, placebo controlled trial in Korea. *Ann. Nutr. Metab.* 66:141–8.

Pascoe, M.C., Crewther, S.G., Carey, L.M., et al. 2011. What you eat is what you are—a role for polyunsaturated fatty acids in neuroinflammation induced depression? *Clin. Nutr.* 30:407–15.

Patrick, R.P., Ames, B.N. 2015. Vitamin D and the omega-3 fatty acids control serotonin synthesis and action, part 2: Relevance for ADHD, bipolar disorder, schizophrenia, and impulsive behavior. *FASEB J.* 29:2207–22.

Peet, M., Laugharne, J., Rangarajan, N., et al. 1995. Depleted red cell membrane essential fatty acids in drug-treated schizophrenic patients. *J. Psychiatr. Res.* 29:227–32.

Pomponi, M., Janiri, L., La Torre, G., et al. 2013. Plasma levels of n-3 fatty acids in bipolar patients: Deficit restricted to DHA. *J. Psychiatric Res.* 47:337–42.

Post, E. 1962. *The significance of affective symptoms in old age.* Institute of Psychiatry. Maudsley Monographs No. 10. London: Oxford University Press.

Pozzi, F., Troisi, A., Cerilli, M., et al. 2003. Serum cholesterol and impulsivity in a large sample of healthy young men. *Psychiatry Res.* 120:239–45.

Poudel-Tandukar, K., Nanri, A., Iwasaki, M., et al. 2011. Long chain n-3 fatty acids intake, fish consumption and suicide in a cohort of Japanese men and women—the Japan Public Health Centerbased (JPHC) prospective study. *J. Affect. Disord.* 129:282–8.

Puri, B.K., Martins, J.G. 2014. Which polyunsaturated fatty acids are active in children with attention-deficit hyperactivity disorder receiving PUFA supplementation? A fatty acid validated meta-regression analysis of randomized controlled trials. *Prostaglandins Leukot. Essent. Fatty Acids* 90:179–89.

Radloff, L.S. 1977. The CES-D scale: A self report depression scale for research in the general population. *Appl. Psychol. Meas.* 1:385–401.

Raine, A., Portnoy, J., Liu, J., et al. 2015. Reduction in behavior problems with omega-3 supplementation in children aged 8–16 years: A randomized, double-blind, placebo-controlled, stratified, parallel-group trial. *J. Child Psychol. Psychiatry* 56:509–20.

Repo-Tiihonen, E., Halonen, P., Tiihonen, J., et al. 2002. Total serum cholesterol level, violent criminal offences, suicidal behavior, mortality and the appearance of conduct disorder in Finnish male criminal offenders with antisocial personality disorder. *Eur. Arch. Psychiatry Clin. Neurosci.* 252:8–11.

Roy, A., Virkkunen, M., Linnoila, M. 1987. Reduced central serotonin turnover in a subgroup of alcoholics. *Biol. Psychiatry* 11:173–7.

Rudin, D.O. 1981. The major psychoses and neuroses as omega-3 essential fatty acid deficiency syndrome: Substrate pellagra. *Biol. Psychiatry* 16:837–50.

Saad, K., Abdel-Rahman, A.A., Elserogy, Y.M., et al. 2016. Vitamin D status in autism spectrum disorders and the efficacy of vitamin D supplementation in autistic children. *Nutr. Neurosci.* 19(8):346–51.

Salib, E., Cortina-Borja, M. 2010. An association between month of birth and method of suicide. *Int. J. Psychiatry Clin. Pract.* 14:8–17.

Sarris, J., Mischoulon, D., Schweitzer, I. 2012. Omega-3 for bipolar disorder: Meta-analyses of use in mania and bipolar depression. *J. Clin. Psychiatry* 73:81–6.

Schopler, E., Reichler, R.J., DeVellis, R.F. et al. 1980. Toward objective classification of hildhood autism: Childhood Autism Rating Scale (CARS). *J. Autism Dev. Disord.* 10:91–103.

Schuchardt, J.P., Huss, M., Stauss-Grabo, M., et al. 2012. Significance of long-chain polyunsaturated fatty acids (PUFAs) for the development and behaviour of children. *Eur. J. Pediatr.* 169:149–64.

Sepehrmanesh, Z., Kolahdooz, F., Abedi, F., et al. 2016. Vitamin D supplementation affects the Beck depression inventory, insulin resistance, and biomarkers of oxidative stress in patients with major depressive disorder: A randomized, controlled clinical trial. *J. Nutr.* 146:243–8.

Servan-Schreiber, D. 2003. *Guérir*. Paris: Robert Laffont.

Sethom, M.M., Fares, S., Bouaziz, N., et al. 2010. Polyunsaturated fatty acids deficits are associated with psychotic state and negative symptoms in patients with schizophrenia. *Prostaglandins Leukot. Essent. Fatty Acids* 83:131–6.

Severus, W.E., Ahrens, B., Stoll, A.L. 1999. Omega-3 fatty acids—the missing link? *Arch. Gen. Psychiatry* 56:380–1.

Shamberger, R.J. 2011. Autism rates associated with nutrition and the WIC program. *J. Am. Coll. Nutr.* 30:348–53.

Sher, Y., Lolak, S., Maldonado, J.R. 2010. The impact of depression in heart disease. *Curr. Psychiatry Rep.* 12:255–64.

Shivakumar, V., Kalmady, S.V., Amaresha, A.C., et al. 2015. Serum vitamin D and hippocampal gray matter volume in schizophrenia. *Psychiatry Res.* 233:175–9.

Siksou, M. 2012. *Introduction à la neuropsychologie clinique*. Paris: Dunod.

Smith, R.S. 1991. The macrophage theory of depression. *Med. Hypotheses* 35:298–306.

Smith, G.D., Pekkanen, J. 1992. Should there be a moratorium on the use of cholesterol lowering drugs? *BMJ* 304:431–4.

Spedding, S. 2014. Vitamin D and depression: A systematic review and meta-analysis comparing studies with and without biological flaws. *Nutrients* 6:1501–18.

Steenweg-de Graaff, J.C., Tiemeier, H., Basten, M.G., et al. 2015. Maternal LC-PUFA status during pregnancy and child problem behavior: The Generation R Study. *Pediatr. Res.* 7:489–97.

Steinert, T., Woelfle, M., Gebhardt, R.P. 1999. No correlation of serum cholesterol levels with measures of violence in patients with schizophrenia and non-psychotic disorders. *Eur. Psychiatry* 14:346–8.

Stevens, L.J., Zentall, S.S., Abate, M.L., et al. 1996. Omega-3 fatty acids in boys with behavior, learning, and health problems. *Physiol. Behav.* 59:915–20.

Stevens, L.J., Zentall, S.S., Deck, J.L., et al. 1995. Essential fatty acid metabolism in boys with attention-deficit hyperactivity disorder. *Am. J. Clin. Nutr.* 62:761–8.

Stoll, A.L., Severus, W.E., Freeman, M.P. 1999. Omega-3 fatty acids in bipolar disorder. A preliminary double-blind, placebo-controlled trial. *Arch. Gen. Psychiatry* 56: 407–12.

Stubbs, G., Henley, K., Green, J. 2016. Autism: Will vitamin D supplementation during pregnancy and early childhood reduce the recurrence rate of autism in newborn siblings? *Med Hypothesis* 88:74–8.

Stumpf, W.E. 2012. Vitamin D and the scientific calcium dogma: Understanding the 'Panacea' of the sun. *Eur. J. Clin. Nutr.* 66:1080–1.

Su, K.P., Lai, H.C., Yang, H.T., et al. 2014. Omega-3 fatty acids in the prevention of interferon-alpha induced depression: Results from a randomized, controlled trial. *Biol. Psychiatry* 76:559–66.

Sublette, M.E., Ellis, S.P., Geant, A.L., et al. 2011. Meta-analysis of the effects of eicosapentaenoic acid (EPA) in clinical trials in depression. *J. Clin. Psychiatry* 72:1577–84.

Sublette, M.E., Hibbeln, J.R., Galfalvy, H., et al. 2006. Omega-3 polyunsaturated essential fatty acid status as a predictor of future suicide risk. *Am. J. Psychiatry* 163:1100–2.

Sullivan, S., Wills, A., Lawlor, D., et al. 2013. Prenatal vitamin D status and risk of psychotic experiences at age 18years-a longitudinal birth cohort. *Schizophr. Res.* 148:87–92.

Szatmari, P., Boyle, M., Offord, D.R. 1989. ADDH and conduct disorder: Degree of diagnostic overlap and differences among correlates. *J. Am. Acad. Child Adolesc. Psychiatry* 28:865–72.

Tanskanen, A., Hibbeln, J.R., Hintikka, J., et al. 2001. Fish consumption, depression, and suicidality in a general population. *Arch. Gen. Psychiatry* 58:512–13.

Tariq, M.M., Streeten, E.A., Smith, H.A., et al. 2011. Vitamin D: A potential role in reducing suicide risk? *Int. J. Adolesc. Med. Health* 23:157–65.

Tiemeir, H., Hofman, A., Kiliaan, A.J., et al. 2002. Vitamin E and depressive symptoms are not related. The Rotterdam Study. *J. Affect. Dis.*72:79–83.

Tsai, A.C., Lucas, M., Okereke, O.I., et al. 2014. Suicide mortality in relation to dietary intake of n-3 and n-6 polyunsaturated fatty acids and fish: Equivocal findings from 3 large US cohort studies. *Am. J. Epidemiol.* 179:1458–66.

Tsuchimine, S., Kaneda, A., Yasui-Furukori, N. 2016. Serum x03C9;-3 and x03C9;-6 Fatty acids are not associated with personality traits in healthy Japanese young people. *Neuropsychobiology* 73:249–53.

Umhau, J.C., George, D.T., Heaney, R.P., et al. 2013. Low vitamin D status and suicide: A case-control study of active duty military service members. *PLoS One* 8:e51543.

Vaisman, N., Kaysar, N., Zaruk-Adasha, Y., et al. 2008. Correlation between changes in blood fatty acid composition and visual sustained attention performance in children with inattention: Effect of dietary n-3 fatty acids containing phospholipids. *Am. J. Clin. Nutr.* 87:1170–80.

Vancassel, S., Durand, G., Barthélémy, C., et al. 2001. Plasma fatty acid levels in autistic children. *Prostaglandins Leukot. Essent. Fatty Acids* 65:1–7.

Vevera, J., Zukov, I., Morcinek, T., et al. 2003. Cholesterol concentrations in violent and nonviolent women suicide attempters. *Eur. Psychiatry* 18:23–7.

Vieth, R., Kimball, S., Hu, A., et al. 2004. Randomized comparison of the effects of the vitamin D3 adequate intake versus 100 mcg (4000 UI) per day on biochemical responses and the wellbeing of patients. *Nutr. J.* 3:8.

Virkkunen, M., Nuutila, A., Goodwin, F.K., et al. 1987a. Cerebrospinal fluid monoamine metabolite levels in male arsonists. *Arch. Gen. Psychiatry* 44:241–7.

Virkkunen, M.E., Horrobin, D.F., Jenkins, D.K., et al. 1987b. Plasma phospholipid essential fatty acids and prostaglandins in alcoholic, habitually violent, and impulsive offenders. *Biol. Psychiatry* 22:1087–96.

Virkkunen, M., Penttinen, H. 1984. Serum cholesterol in aggressive conduct disorder: A preliminary study. *Biol. Psychiatry* 19:435–9.

Von Schacky, C. 2010. Omega-3 Index and cardiovascular disease prevention: Principle and rationale. *Lipid Technol.* 22:151–4.

Vuksan-Cusa, B., Marcinko, D., Nad, S., et al. 2009. Differences in cholesterol and metabolic syndrome between bipolar disorder men with and without suicide attempts. *Prog. NeuroPsychopharmacol. Biol. Psychiatry* 33:109–12.

Wallner, B., Machatschke, I.H. 2009. The evolution of violence in men: The function of central cholesterol and serotonin. *Prog. NeuroPsychopharmacol. Biol. Psychiatry* 33:391–7.

Weidner, G., Connor, S.L., Hollis, J.F., et al. 1992. Improvements in hostility and depression in relation to dietary change and cholesterol lowering. The Family Heart Study. *Ann. Intern. Med.* 117:820–3.

Williams, R.B., Littman, A.B. 1996. Psychosocial factors: Role in cardiac risk and treatment strategies. *Cardiol. Clin.* 14:97–104.

Wittchen, H.U., Jacobi, F., Rehm, J., et al. 2011. The size and burden of mental disorders and other disorders of the brain in Europe 2010. *Eur. Neuropsychopharmacol.* 21:655–79.

Wozniak, J., Biederman, J., Mick, E., et al. 2007. Omega-3 fatty acid monotherapy for pediatric bipolar disorder: A prospective open-label trial. *Eur. Neuropsychopharmacol.* 17:440–7.

Wu, S., Ding, Y., Wu, F., et al. 2015. Serum lipid levels and suicidality: A meta-analysis of 65 epidemiological studies. *J. Psychiatry Neurosci.* 41:150079.

Wunderlich, U., Bronisch, T., Wittchen, H.U. 1997. Comorbidity patterns in adolescents and young adults with suicide attempts. *Eur. Arch. Psychiatry Clin. Neurosci.* 248:87–95.

Yoshikawa, E., Nishi, D., Matsuoka, Y. 2015. Fish consumption and resilience to depression in Japanese company workers: A cross-sectional study. *Lipids Health Dis.* 14:51.

Yui, K., Koshiba, M., Nakamura, S., et al. 2012. Effects of large doses of arachidonic acid added to docosahexaenoic acid on social impairment in individuals with autism spectrum disorders: A double-blind, placebo-controlled, randomized trial. *J. Clin. Psychopharmacol.* 32:200–6.

Zaalberg, A. 2015. Nutrition, neurotoxicants & aggressive behaviour. PhD Diss. Nijmegen. http://www.repository.ubn.ru.nl/handle/2066/141039.

Zaalberg, A., Nijman, H., Bulten, E., et al. 2010. Effects of nutritional supplements on aggression, rule-breaking, and psychopathology among young adult prisoners. *Aggr. Behav.* 36:117–26.

Zaparoli, J.X., Galduróz, J.C. 2012. Treatment for tobacco smoking: A new alternative? *Med. Hypotheses* 79:867–8.

Zhang, Y., Leung, D.Y., Richers, B.N., et al. 2012. Vitamin D inhibits monocyte/macrophage proinflammatory cytokine production by targeting MAPK phosphatase-1. *J. Immunol.* 188:2127–35.

chapter seven

Annexes

7.1 Essential fatty acids

The ω-3 fatty acids are biochemically different from their counterparts, the ω-6 fatty acids, by the position of their first double bond, situated between the third and the fourth carbon atom instead of being between the sixth and the seventh carbon atom. According to that nomenclature, the carbon numbering starts from the terminal methyl group (opposite to the acid function), hence their name ω-3 (or n-3) and ω-6 (or n-6). Only plants have the ability to convert linoleic acid ([LA] 18:2 ω-6) (Figure 7.1), the precursor of the ω-6 fatty acid series, into α-linolenic acid ([ALA] 18:3 ω-3) (Figure 7.2), the precursor of all the ω-3 fatty acid series.

Animals (including humans) do not have the capacity to synthesize these two fatty acids (considered as essential) and their supply can be done only from food. In humans as in animals, the same enzymes along the seven metabolic steps are involved in the biosynthesis of long-chain ω-3 fatty acids, as eicosapentaenoic acid (EPA), docosahexaenoic acid (DHA), or of long-chain ω-6 fatty acids, as arachidonic acid (AA), with each family being not convertible into one other (Table 7.1). This property results in competition for the biosynthesis of the products from LA and ALA, as well as to their metabolic derivatives (prostaglandins and various oxygenated or hydroxy derivatives). Thus, an excess of LA may reduce the production of EPA and DHA from ALA, the latter of which is present in very low amounts in animal tissues.

The ω-3 fatty acid content of biological samples (red blood cells, whole blood, plasma) is now often estimated by calculating the "ω-3 index." The value of this index, calculated as the sum of EPA plus DHA in percentage of the total fatty acids in the sample, provides an indication of the ω-3 fatty acid status for the whole body. High blood EPA plus DHA levels (>8%) are found in people from regions near the Sea of Japan, in Scandinavia, and in areas with indigenous populations or populations not fully adapted to westernized food habits. Very low blood levels (<4%) were observed in people of North America, Central and South Americas, Europe, the Middle East, Southeast Asia, and Africa.

It is generally estimated that an ω-3 index of 8% or more is an ideal goal for general good health. If the value of the index is 4% or less, it is recommended to change food habits by consuming more marine animals (fish, molluscs, shellfish) or by ingesting food supplements rich in long-chain

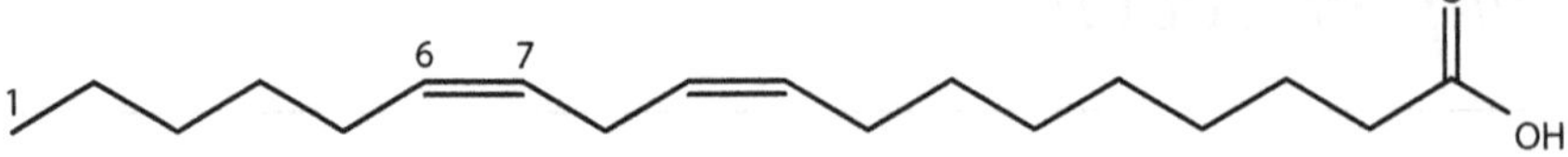

Figure 7.1 Linoleic acid (18:2 ω-6, LA).

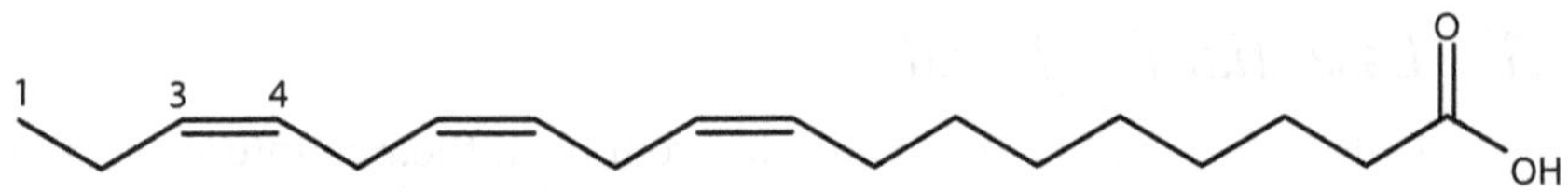

Figure 7.2 Linolenic acid (18:3 ω-3, ALA).

ω-3 fatty acids (EPA, DHA, or both). In that case, the generally admitted practice is to use a pharmaceutical-quality product supplying 500–1000 mg of EPA and DHA per day. To optimize the dietary ω-3 fatty acid intake, it is necessary to look at a precise table of food fatty acid composition. A convenient table containing 8000 different foods with their ω-3 and ω-6 fatty acid composition and the value of their ω-3 index may be consulted on a dedicated website (www.fattyacidshub.com/tools-for-fatty-acids/omega-3-omega-6-ratio-calculator/). From a practical point of view, Prof. Bill Lands has established the "Omega 3-6 Balance Scores" for nearly 5000 foods, providing a simple way to highlight the essential fatty acid state in a diet and allowing to plan more accurately better balanced meals (www.efaeducation.org/Omega3-6BalanceApp.html).

7.2 Dietary allowance of essential fatty acids

The tables (Tables 7.2–7.4) below summarize the opinion of the French Food Safety Agency (AFSSA) concerning the updating of the recommended dietary allowances for fatty acids (saisine no. 2006-SA-0359, 1 March 2010). Similar data may be found in various official or medical sources from several countries.

7.2.1 Main DHA and EPA sources as sea products

Despite heavy metals and pesticides pollution, fish remain the most convenient and most abundant source of long-chain ω-3 fatty acids. In terms of risk, in 2014 two US agencies (Food and Drug Administration, Environmental Protection Agency) recommended for pregnant women or those planning to become pregnant to not eat more than three servings of fish per week to limit the exposure of the fetus to mercury. Sweden recommended limiting the consumption to once per month. The UK authorities

Table 7.1 Biosynthetic pathways of ω-6 and ω-3 fatty acids (enzymes responsible are in italics). The first digit indicates the number of carbon atoms, the second the number of double bonds

Metabolic step	ω-6 Fatty acid	Enzyme	ω-3 Fatty acid
	18:2 ω-6 (LA) Linoleic acid		18:3 ω-3 (ALA) Linolenic acid
1		*Δ6 desaturase*	
	18:3 ω-6 γ-Linolenic acid		18:4 ω-3 Stearidonic acid
2		*elongase*	
	20:3 ω-6 Dihomo-γ-linolenic acid		20:4 ω-3 Eicosatetraenoic acid
3		*Δ5 desaturase*	
	20:4 ω-6(AA) Arachidonic acid		20:5 ω-3 (EPA) Eicosapentaenoic acid
4		*elongase*	
	22:4 ω-6 Docosatetraenoic acid		22:5 ω-3 Docosapentaenoic acid
5		*elongase*	
	24:4 ω-6 Tetracosatetraenoic acid		24:5 ω-3 Tetracosapentaenoic acid
6		*Δ6 desaturase*	
	24:5 ω-6 Tetracosapentaenoic acid		24:6 ω-3 Tetracosahexaenoic acid
7		*beta-oxidation*	
	22:5 ω-6 Docosapentaenoic acid		22:6 ω-3 (DHA) Docosahexaenoic acid

Table 7.2 Recommended dietary intakes of polyunsaturated fatty acids (as % of the total energy intake or mg) for pregnant women and nursing mothers, consuming respectively 2050 and 2250 kcal (or 8583 and 9420 kJ) with 35%–40% of energy intake as lipids

	Linoleic acid 18:2 ω-6 (LA), %	Linolenic acid 18:3 ω-3 (ALA), %	Arachidonic acid 20:4 ω-6 (AA)	Docosahexaenoic acid 22:6 ω-3 (DHA), mg	EPA+DHA, mg
Pregnant women	4	1	–	250	500
Nursing mothers	4	1	–	250	500

Table 7.3 Recommended dietary intakes of polyunsaturated fatty acids for newborn or infant (in % of total energy intake or % of total fatty acids, TFA)

	Linoleic acid 18:2 ω-6 (LA), %	Linolenic acid 18:3 ω-3 (ALA), %	Arachidonic acid 20:4 ω-6 (AA), %	Docosahexaenoic acid 22:6 ω-3 (DHA)	EPA+DHA
Newborn or infant	2.7	0.45	0.5	0.32 TFA	EPA<DHA

Table 7.4 Recommended dietary intakes of polyunsaturated fatty acids for infants aged of more than 6 months, children, and adolescents (in % of total energy intake or mg)

	Linoleic acid 18:2 ω-6 (LA), %	Linolenic acid 18:3 ω-3 (ALA), %	Arachidonic acid 20:4 ω-6 (AA)	Docosahexaenoic acid 22:6 ω-3 (DHA), mg	EPA +DHA, mg
Infants aged 0-1	2.7	0.45	–	70	–
Infants aged 1-3	2.7	0.45	–	70	–
Infants aged 3-9	4	1	–	125	250
Infants aged 10-18	4	1	–	250	500

recommend pregnant women avoid eating certain marine fish (tuna, shark, marlin, swordfish, bar). In 2004 in France, the AFSSA recommended that pregnant and lactating women, and children aged up to 30 months, consume no more than 60–150 g/week of these wild predatory fish.

New studies on mercury contamination of fish consumed in Europe are alerting consumers about safe versus unsafe fish consumption (see below). To regularly eat fish is thus not without risk to health.

Among fatty fish, some contain more DHA and EPA that others and are therefore particularly interesting for a balanced diet.

Fish may be grouped into three categories:

- Fatty fish with high ω-3 amounts (>1.5 g/100 g): salmon, mackerel, herring
- Fatty fish with medium ω-3 amounts (0.5–1.5 g/100 g): trout, sardine, tuna
- Lean fish with low ω-3 amounts (<0.5 g/100 g): sole, cod, carp, eel

Table 7.5 shows that even a modest consumption of marine fatty fish, or even farmed salmon or trout, is able to cover the ω-3 fatty acid requirements

Table 7.5 EPA and DHA content (g/100 g, in descending order) of various fish, shellfish and molluscs and weight of a portion providing 500 mg of these fatty acids

Fish	EPA+DHA (g/100 g)	Fish weight (g per 500 mg EPA+DHA)
Farmed salmon	3.1	16
Mackerel	2.3	22
Salmon	1.96	26
Herring	1.57	32
Sardine	0.98	51
White tuna	0.86	58
Farmed trout	0.74	68
Shrimp	0.55	90
Mussel	0.44	114
Oyster	0.39	128
Octopus	0.31	161
Sole	0.25	200
Cod	0.18	278

of an adult. A 150- to 200-g portion of some of these fish (right column), consumed two times per week, covers the requirements of an adult for EPA and DHA (about 500 mg/day).

7.2.2 *Foods for infants and young children*

Since 1996, an amendment to the European Union Directive on infant formulas allowed the marketing of milk enriched with polyunsaturated fatty acids, specifying the limits to be respected, but without setting a minimum threshold for DHA (and AA). For specialists, an enrichment corresponding to the amounts found in breast milk seems the most appropriate. Many foods (milk) supplemented with polyunsaturated fatty acids for infants or children are already marketed in various countries. Some of them are detailed below:

- Bon départ from Nestlé contains 10 mg of DHA and 56 mg of LA in 100 mL of preparation.
- Enfamil A + from Mead Johnson contains 11.5 mg of DHA in a 100-mL preparation.
- Candia source d'Omega-3 from Candia is a milk containing 1.7 g of lipids with 23 mg of EPA and 33 mg of DHA per 100 mL.
- Calisma 2 from Gallia contains 8.3 mg of DHA and the product "croissance" 12 mg of DHA in 100 mL of preparation.

For other products, consumers must check the manufacturers' indications on the fatty acid composition and especially the presence of DHA and EPA in favoring formulas containing low amounts of ω-6 fatty acids.

7.2.3 Food supplements rich in EPA, DHA, or both

Because the recommended dietary allowance for ω-3 fatty acids is about 500 mg/day, usually in the form of a mixture of EPA and DHA, a supplementation may be provided by the daily intake of these fatty acids in capsules (0.5 or 1 g). They contain esterified derivatives of free fatty acids (usually as ethyl esters) or better fish or shellfish (krill) oil (triacylglycerols). These oils are generally declared to be low in heavy metals and pesticides. They are also enriched in vitamin E to ensure a better conservation (usually about 10 mg per capsule). These oils are extracted from cold water fish (anchovies, sardine, cod, salmon, hoki), from Antarctic water shellfish (krill), and even from marine mammals (seals). They are used after purification, usually by molecular distillation to remove heavy metals and pesticides. Their ω-3 fatty acid concentration ranges from 30% to >90%. Krill oil is characterized by its very low level of organic pollutants and heavy metals, thereby justifying the intense development of its production.

Now, the ω-3 fatty acids on the market are mainly extracted from fish and to a lesser extent from shellfish and algae. It is expected that the growing worldwide demand for DHA and EPA will push manufacturers to offer new products derived from seaweed and possibly from genetically modified plants.

Among oils from marine animals and algae present on the market, one can find the following brands:

- Arctic Omega-3 seal oil with 500 mg of Arctic seal oil containing 120 mg of ω-3 fatty acids, including 36 mg of EPA, 51 mg of DHA, and 31 mg of 22:5 ω-3
- DHA Neuromins-Solgar with 500 mg of oil containing 100 mg of DHA from seaweed
- Fitoform DHA omega-3 vegetal, 1-g capsule containing 12 mg of EPA, 152 mg of DHA, and 12 mg of LA
- CNG Salmon oil, 1 g of oil containing 180 mg of EPA and 120 mg of DHA
- NKO Krill, 300 mg of Antarctic shrimp oil, mainly phospholipids containing 39 mg of EPA and 16.5 mg of DHA (Oemine brand: a capsule of 500 mg NKO krill oil contains 75 mg of EPA, 45 mg of DHA, and 0.75 mg of astaxanthin)
- Marin EPA Platinum (Nutrogenics), capsule of purified oil containing 764 mg of EPA, 236 mg of DHA, and 300 international units (IU) of vitamin D

- Omacor soft caps, 1 g of ω-3 fatty acid ethyl esters with 460 mg of EPA and 380 mg of DHA
- Omega-Brite, 500 mg of 90% ω-3 fatty acids with EPA (350 mg) and DHA (50 mg)
- Omegavie DHA 1050 TG/EE (Polaris nutritional lipids), capsule containing 7%–17% EPA and 50% DHA
- Triglistab, 1010-mg capsule containing 700 mg of ω-3 fatty acid ethyl esters
- Ultimate Omega, 1 g of anchovy and sardine oil containing 640 mg of ω-3 fatty acids as EPA (325 mg) and DHA (225 mg)
- Unocardio (WHC-Nutrogenics), capsule containing 460 mg of EPA and 380 mg of DHA
- Xtend-Life Omega/DHA Fish Oil (hoki oil), 1-g capsule with 120 mg of EPA, 280 mg of DHA, and 50 mg of 22:5 ω-3.
- Ysomega, 1 g of fish oil containing 320 mg of EPA and 200 mg of DHA
- ZenixX (IXX Pharma), 1-g capsule containing 60%–75% of DHA and 15% EPA ethyl esters

The reader may explore a comprehensive list of fish oil products with all the testing results established by the International Fish Oil Standards program (www.nutrasource.ca/ifos/product-reports/default.aspx).

For problems of fish mercury contamination, see the 2010 report of the joint Food and Agriculture Organization of the United Nations/World Health Organization (WHO) expert consultation on the risks and benefits of fish consumption (http://www.fao.org/docrep/014/ba0136e/ba0136e00.pdf) and the 2015 EFSA's official publication (http://onlinelibrary.wiley.com/doi/10.2903/j.efsa.2015.3982/pdf) (report of the Scientific Committee "Statement on the benefits of fish/seafood consumption compared to the risks of methylmercury in fish/seafood").

Unexpected solutions could result from studies performed in animals in 2012 by the team of Dr. C. Stanton, University of Cork, Ireland, showing that the ingestion of certain strains of *Bifidobacterium* could increase the brain DHA content. After a possible verification in humans, this effect could act synergistically with a moderate food supplementation of ω-3 fatty acids.

7.3 Vitamin A and carotenoids

7.3.1 Vitamin A

Vitamin A (Figure 7.3) is a generic term for several lipidic compounds related to retinol and having the same biological activity, such as retinol esters, retinal, and retinoic acid. They all belong to the group of retinoids. They are formed from a carotenoid by splitting the middle of the carbon chain.

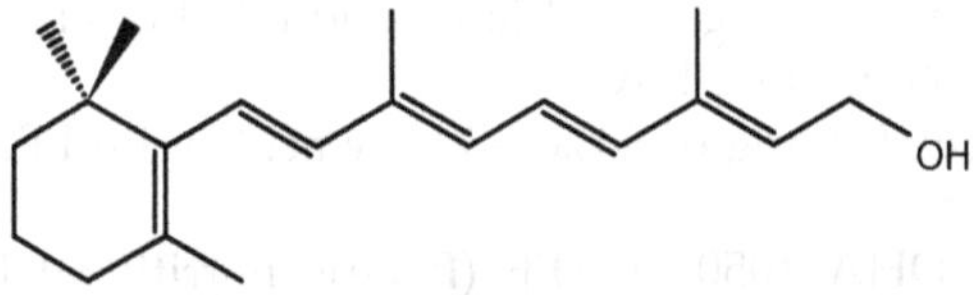

Figure 7.3 Vitamin A (all-*trans*-retinol).

In animals, they offer two vitamin activities: photoreception (retinol, retinal) and cell growth regulation (retinoic acid).

For humans, natural sources of vitamin A are retinoids included in consumed animal products (meat, eggs, dairy products) and formed from some carotenoids (provitamin A) of plant origin.

The amounts of vitamin A are expressed in micrograms of retinol equivalent (RE), with that unit enabling the conversion of all sources of vitamin A in a single expression. Thus, it is recognized that 1 RE is equivalent to 1 μg of retinol, 1.78 μg of retinyl palmitate, and 6 μg of β-carotene.

WHO has recommended humans have a dietary allowance of 600 μg RE. In France, for adults, AFSSA has set daily vitamin A intake to 800 μg RE. A partial deficiency status has been highlighted in many countries. Indeed, the WHO estimates that several millions of children are deficient worldwide. This situation leads nearly 500,000 children to blindness each year in developing countries, mainly South Asia and Africa. It can be also considered that the French population is currently in a deficiency state because it was established that about 72% of a population between 50 and 68 years have a vitamin A intake lower than that recommended. To establish a reliable vitamin A intake, it is recommended to have a diet that includes various fruits and vegetables. These vegetables will be consumed preferably cooked in the presence of a little oil to facilitate carotenoid absorption. The consumption of liver (10,000 RE/100 g) and cheese (200–400 RE/100 g) is also recommended for direct supply of retinol. It is possible to carry out a daily supplementation of about 1 mg of vitamin A, not to exceed 3 mg/day (Leray 2015, *Lipids: Nutrition and Health*, CRC Press, Boca Raton, FL).

7.3.2 *Carotenoids*

β-Carotene is the best known carotenoid (Figure 7.4). It gives some fruits and vegetables an orange-red color, but it is always present hidden by the green chlorophyll in plants. Like all carotenoids, it protects cells against free radicals formed by the action of light energy.

It is used as a food coloring (E160a) and also as provitamin A in vitamin supplements.

The following carotenoids are not provitamin A: lutein (Figure 7.5) and zeaxanthin. Lutein is a carotenoid of the xanthophyll group. It is present

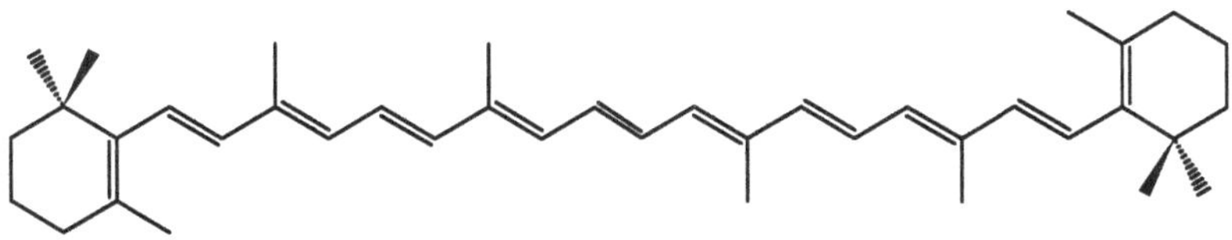

Figure 7.4 β-Carotene.

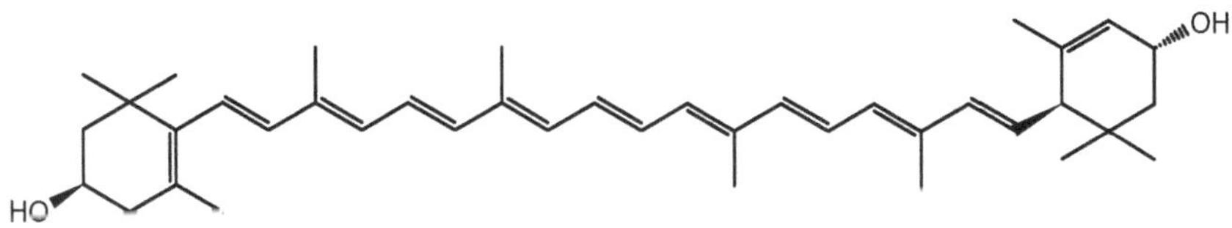

Figure 7.5 Lutein.

in some foods of vegetal origin and characterizes the retinal macula. Zeaxanthin is a *cis* isomer of lutein (both hydroxyl groups in the same plane). Three stereoisomers are present in human serum and in retina, the most abundant being *meso*-zeaxanthin (3R,3′S-zeaxanthin). Lutein (E161b) and zeaxanthin (E161h) are permitted as food additives.

The main dietary sources of lutein are vegetables with green leaves, in descending order (from 15 to 1 mg/100 g), cabbage, watercress, spinach, chard, small peas, broccoli, and lettuce. Fruits such as oranges and peaches are also significant sources of zeaxanthin.

It has been determined that dietary fortification with macular carotenoids (10 mg of lutein, 2 mg of zeaxanthin, and 10 mg of *meso*-zeaxanthin) can have meaningful effects on visual function.

In France, the AFSSA considers that there is no scientific evidence to justify a lutein or zeaxanthin supplementation in healthy humans with a varied diet. However, it is possible to use a supplement of 6 mg/day of lutein, alone or combined with zeaxanthin. Some health professionals also recommend to their patients such a supplementation. Higher doses (10–20 mg) are prescribed for the prevention of cataract and age-related macular degeneration (AMD) in individuals at risk (Leray 2015).

7.4 Vitamin D

This vitamin is represented by several molecules derived from sterols (cholesterol in animals and ergosterol in plants) by solar ultraviolet irradiation (UV-B) and by enzymatic hydroxylation. The most abundant form is the cholecalciferol (vitamin D_3, Figure 7.6), which can be considered a

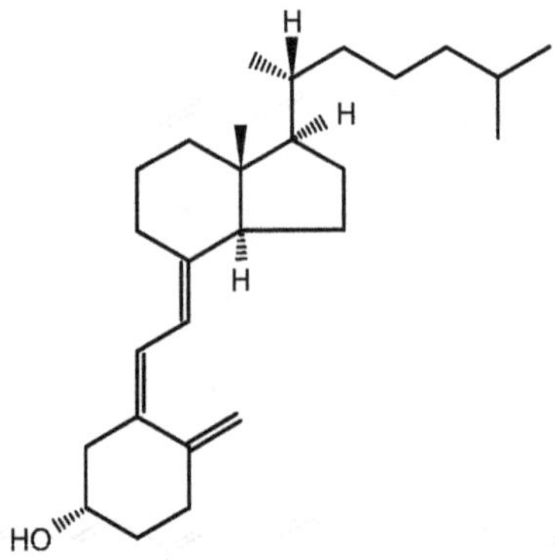

Figure 7.6 Vitamin D_3 (cholecalciferol).

provitamin D. It is metabolized when required in the liver into a circulating form, 25-hydroxycholecalciferol (calcidiol), and then in the kidney into the active form of the vitamin, 1,25-hydroxycholecalciferol (calcitriol). Considering its biosynthesis pathway and its mode of action, that vitamin is now frequently taken as a hormone. It is sometimes called "solar hormone."

The amounts of vitamin D are expressed in milligrams (or in nanograms) of cholecalciferol. The IU may also be used, with 40 IU being equivalent to 1 μg of vitamin D. Blood concentrations are expressed in nanograms per milliliter or nanomoles per liter (1 ng/mL = 2.6 nmol/L).

Initially discovered as part of the regulation of calcium and phosphate metabolism, vitamin D is now recognized as having a crucial role in metabolic disorders, diabetes, cardiovascular diseases, cancers, immune system, and brain function (Leray 2015).

Vitamin D is a lipid present in some foods, such as fatty fish (herring, sardine, tuna, salmon, and mackerel), milk, and eggs, but its supply is mainly via synthesis in the skin under the effect of sun exposure. Indeed, it has been shown that fish consumption in accordance with the recommended amounts is able to increase the blood vitamin D level, but it cannot alone optimize its status.

UV irradiation of ergosterol, contained in vegetable oils and fungi, produces ergocalciferol (vitamin D_2). Its biological activities with respect to vitamin D_3 have not yet been precisely determined; for some they are three times lower, whereas for the others the two forms are equivalent. Ergocalciferol is poorly represented in the diet, but it is often provided by pharmaceutical preparations used as vitamin D supplement.

As recalled Dr. J.-C. Souberbielle, Hospital Necker-Enfants-Malades, Paris, France: "from 43% to 50% of the adult French population ('healthy') have a vitamin D concentration less than 20 ng/mL (deficiency state) and more than 80% have a concentration below 30 ng/mL (insufficient state between 20 and 30 ng/mL). These percentages are even higher in subjects aged over 70–80 living at home and among teenagers in apparent good health, and increase further in patients with various pathologies."

It is estimated that more than 1 billion people worldwide are vitamin D deficient, with the intake in Western countries often being less than 9 µg/day. Deficiency states are more common in women than in men, with pregnancy representing a situation at risk. This situation can be aggravated by obesity, age, season, a highly pigmented skin, or a full body covering. There is a possibility that the patients treated with statins may not have enough cholesterol because it is effectively the primary substrate for vitamin D production.

It is possible to assess the risk of vitamin D deficiency by using a questionnaire to determine the contribution of sun exposure and that of the food (Diagno® vitamin D, www.nutritionpreventiveisio.fr).

The prevention of the vitamin D deficiency in Europe is a subject of such a priority that the European Commission has decided to support a large multidisciplinary research project, the ODIN project, in 19 countries to prevent that deficiency and improve the public health through nutrition (http://www.odin-vitd.eu). This 4-year project should provide reliable results by the end of 2018.

New recommendations for daily intakes of vitamin D were put forward by health authorities in various countries. Thus, the daily requirements of vitamin D are at least as follows:

- From 0 to 1 year old: 10 µg (400 IU) per day
- From 1 year to 70 years old (including pregnant women): 15 µg (600 IU) per day
- After 70 years old: 20 µg (800 IU) per day

Many experts have claimed that this is still far too low to address chronic insufficiency of the vitamin that comes with modern life wherein people spend much of their time indoors and use sunscreen when they are outdoors.

The European Food Safety Authority (EFSA) for the first time (March 2016) issued vitamin D intake recommendations for European adults, pregnant women, children, and infants. After consideration of factors such as sun exposure, the panel set an adequate intake level of 15 µg/day from food sources for adults and children to achieve a serum level of 19 ng/mL (50 nmol/L). For infants aged 7–11 months, a supply of 10 µg/day was established.

In July 2016, the Public Health of England (Scientific Advisory Committee on Nutrition) advised that without taking into account sunlight exposure, adults and children in the United Kingdom over the age of 1 year should get 10 μg of vitamin D every day. People who have a higher risk of vitamin D deficiency are being advised to take a supplement year-round. As a precaution, babies aged less than 1 year should have a daily 8.5–10 μg vitamin D supplement, whereas children aged 1–4 years should have a daily 10 μg of vitamin D supplement year-round.

The French government authorities decided early on that it was necessary to fight against vitamin D deficiency in children, deciding a prophylaxis was mandatory for all infants from 0 to 18 months; then in winter, in children up to 5 years old (ministerial circulars, February 21, 1963 and January 6, 1971). The prophylaxis is recommended in children and adolescents when the life conditions provide an insufficient sunlight exposure or a diet low in vitamin D.

According to the French Academy of Medicine, the currently recommended daily intakes are not enough; they should be doubled to reach 30 μg per adult. Since 2012, the academy recommends that the vitamin status of an individual must be defined by the calcidiol serum level, with the latter being higher than 30 ng/mL (75 nmol/L) without exceeding 200 nmol/L. Any vitamin D deficiency should be corrected only by an oral supplementation and should not lead subjects to extend their sun exposure. The supplementation will be in the form of a regular intake of vitamin D, corresponding to approximately 800 IU from 1 to 8 years old, 1000 IU from 9 to 50 years old, 1500 IU from 51 to 70 years old, and more than 1500 IU after the age 70 years. Practically, that supply is done each month or every 2 months (e.g., with 100,000 IU per ampoule).

Because most vitamin D experts agree that an intake lower than 10,000 IU/day is safe in healthy adults, the US Endocrine Society has recommended 1500–2000 IU/day as the minimum dose.

In France, short sunlight exposure may cover the vitamin D needs: about 15 min of exposure (hands, forearms, and face), two to three times a week between 9 am and 16 pm from spring to autumn. Notably, vitamin D synthesis cannot be carried out through a window or in the presence of sunscreen.

Natural sources of vitamin D:

- Canned Cod Liver: 100-g supply 54 μg of vitamin D (250 μg for 100 g of cod liver oil).
- Salmon: pink salmon has two times more vitamin D (15 μg/100 g) than wild Atlantic salmon (8 μg/100 g), with the latter having slightly more vitamin D when produced by farming (7 μg/100 g). Trout is a little less rich in vitamin D (11 μg/100 g).
- Herring: 100 g is sufficient to supply almost half the daily requirements. The cooking mode influences the level of vitamin D: 100 g of marinated herring supplies 12–22 μg of vitamin D, when oven cooked or broiled a piece of 100 g supplies 16 μg.

- Fried mackerel, grilled or canned sardines: 100 g provides about 12 μg of vitamin D.
- Canned tuna: 100 g of natural white tuna (albacore) contains about 5 μg of vitamin D; 100 g of tuna in oil contains 3 μg.
- Egg yolk: a 100-g raw egg yolk provides 3 μg of vitamin D, but 2 μg/100 g for cooked egg yolk.
- Cow milk: 250 mL of whole milk provide 3 μg of vitamin D, equivalent to 20% of the recommended daily intake in children and adults.
- Veal liver: 100 g provide about 2.5 μg of vitamin D, nearly one quarter of the recommended daily intake. Beef liver is a little less rich (about 1.2 μg).

It is now possible to find on the French market dairy products fortified with vitamin D (milk, yogurt). There are also vegetable oils supplemented with vitamin D (Isio 4 Lesieur with 0.5 mg/L).

7.5 Vitamin E

Vitamin E is a generic term that encompasses eight molecules with closely related structures (vitamers). They may be distributed in two groups: four tocopherols or vitamin E with a saturated side chain (Figure 7.7) and four tocotrienols or vitamin E with an unsaturated side chain (Figure 7.8). All these compounds are specific to the plant world and essential to humans who have to get them from their food (Leray 2015).

α-Tocopherol is the most common form and is often used to define chemically and physiologically the "vitamin E."

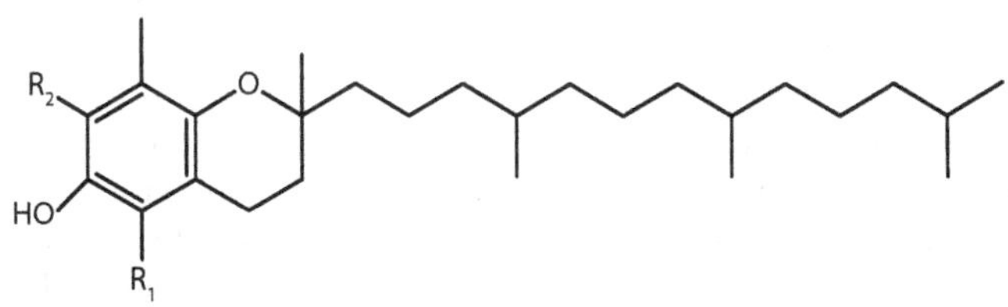

R_1	R_2	
CH_3	CH_3	α-Tocopherol
CH_3	H	β-Tocopherol
H	CH_3	γ-Tocopherol
H	H	δ-Tocopherol

Figure 7.7 Tocopherols.

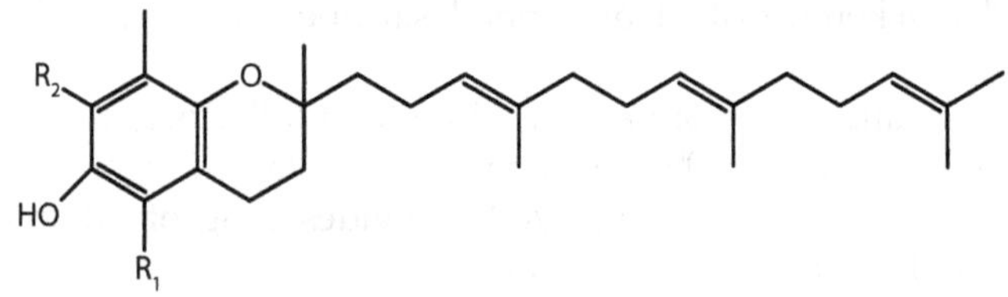

R_1	R_2	
CH_3	CH_3	α-Tocotrienol
CH_3	H	β-Tocotrienol
H	CH_3	γ-Tocotrienol
H	H	δ-Tocotrienol

Figure 7.8 Tocotrienols.

All compounds forming the vitamin E complex are, to varying degrees, powerful antioxidants, protecting cellular and circulating lipids.

The amounts of vitamin E are expressed in milligram equivalents of α-tocopherol; the other vitamers are usually estimated by applying a specific coefficient (0.56 for β-tocopherol, 0.16 γ-tocopherol, 0.5 for δ-tocopherol, and 0.16 for α-tocotrienol).

The IU is not frequently used, with 1 IU corresponding to 0.67 mg of natural α-tocopherol (RRR-α-tocopherol) or 0.91 of synthetic α-tocopherol (*all-rac*-α-tocopherol).

In the United States, the recommended daily allowance is 15 mg of α-tocopherol in adults for both men and women. A serum α-tocopherol concentration of ≥30 μmol/L is considered as a desirable target for health benefits, with manifest clinical symptoms of vitamin deficiency being reported below 8 μmol/L. It has been estimated that in the United Kingdom and the United States more than 75% of the population do not meet the recommended daily intake.

The EFSA recommends a daily intake of 5 mg vitamin E (α-tocopherol equivalent) in the very young child (7–11 months) and an intake from 6 to 9 mg in children (1–10 years). In adolescents (>10 years) and adults, the EFSA recommends an intake of 11 mg in women and 13 mg in men. For people of more than 75 years, the recommended intake should be higher (up to 50 mg/day). In pregnant or nursing women, the EFSA considers that it is not necessary to increase the vitamin E dietary intake.

Plant products are the main source of vitamin E in humans, with the richest being oils from seeds and grains (maximum for wheat germ oil, 2600 mg/L). The distribution of the different vitamin E vitamers varies with the source: wheat, corn, soybean, and peanut contain mostly tocopherols, whereas palm oil, rice, oat, and barley contain mostly tocotrienols. The vitamin E content of fruits and vegetables is very low. Thus, the vitamin E status may be improved in encouraging the consumption of vitamin E-rich food sources (e.g., vegetables, dairy products, eggs), an adequate fortification of food products (e.g., vegetable oils), and a supplementation.

7.6 Cholesterol

Cholesterol (Figure 7.9), as phytosterols or sterols, belongs to the large group of steroids. All derive from squalene, after formation of a core with four cycles, the sterane core. Cholesterol may be the source of many other compounds, such as steroid hormones, bile acids, and vitamin D.

Cholesterol is the major sterol in animals where it participates in building cell membranes to which it brings a certain rigidity. It is particularly concentrated in the nervous system, adrenal glands, liver, and gallstones. It was isolated for the first time in 1770 by F. Poulletier de la Salle in gallstones and found in animal fats in 1815 by M. E. Chevreul, who named it "cholesterine" (from the Greek khole = bile and stereos = solid). With few exceptions, cholesterol is absent from plants.

It is estimated that the body of an adult man contains about 100 g of cholesterol and that his daily needs are about 1 g. Ten percent to 20% of that supply comes from food, with the rest being produced endogenously (mainly in the liver and intestines). Cholesterol metabolism varies among individuals and probably depends on their age. Disruption of its biosynthesis is the cause of hypercholesterolemia, whether hereditary or acquired. The regulation of cholesterol biosynthesis occurs mainly in the liver from acetyl-coenzyme A (CoA) through several enzymatic steps and more particularly through the key enzyme in that metabolism, hydroxymethylglutaryl-CoA reductase. This enzyme is also the target of statins, a family of drugs used to fight against elevated circulating cholesterol. When excessive cholesterol amounts are absorbed from the intestine, its synthesis in the liver is less efficient. It follows that, in the majority of subjects, any dietary intervention will have little impact on blood cholesterol levels.

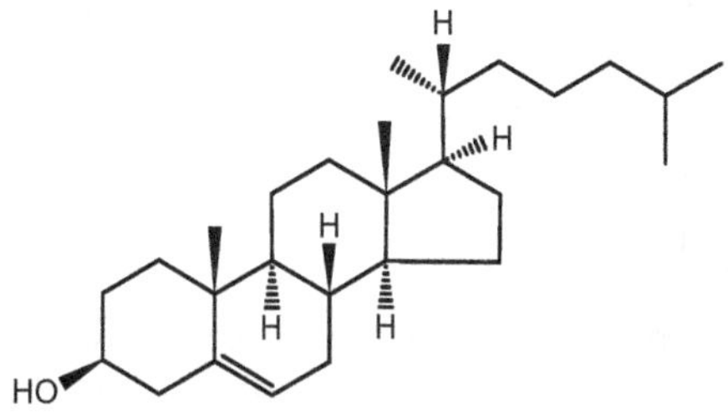

Figure 7.9 Cholesterol.

There is no official recommendation for a daily cholesterol intake. Although a positive link between dietary cholesterol and a risk of cardiovascular disease is suspected, an important intake reduction does not seem desirable because it would be necessarily accompanied by a large change in eating habits. It seems however reasonable from a dietary point of view not to exceed an intake of 300 mg of cholesterol per day. It is recognized that the plasma concentration of total cholesterol in healthy adults should be between 1.8 and 2.5 g/L (between 4.6 and 6.3 mmol/L). Among the commonly consumed foods, only five deserve a special attention: three offals (brain, kidney, and liver), eggs, and butter. Because of their high cholesterol content, it is recommended that they be moderately consumed (Leray 2015).

7.7 Phospholipids

Phospholipids are mostly glycerophospholipids; therefore, they contain one molecule of glycerol esterified with two fatty acids, most frequently different from each other, and a "head group." This polar head group gives an originality to phospholipids; thus, for the most important, it is a phosphocholine for phosphatidylcholine, a phosphoserine for phosphatidylserine, and a phosphoethanolamine for phosphatidylethanolamine. They are the main lipid constituents of cellular membranes and are therefore present in all meats, but they also are in the form of fat depots in milk and egg yolk. Plants also contain these phospholipids, but their content remains low (up to 0.2%), greatly reducing their nutritional value except when consumed in the form of supplements prepared from seeds. Their interest lies in their polar head and their fatty acid content (Leray 2015).

7.7.1 Phosphatidylcholine

Phosphatidylcholine (Figure 7.10) is the most abundant phospholipid in animal and plant tissues (almost 50% of total phospholipids). It contains about 14% by weight choline. The most often encountered fatty acid is palmitic acid (16:0) in the *sn*-1 position (R1) and oleic acid (18:1 ω-9) or LA (18:2 ω-6) in the *sn*-2 position (R2).

The historical term "lecithin" (from the Greek lecithos, egg yolk) is often used for this phospholipid. It was named by the French chemist N. T. Gobley in 1845 after its discovery in the egg yolk. By extension, the term lecithin is used in food industry to denote a mixture rich in phospholipids (at least 60%); therefore in phosphatidylcholine, its content in neutral lipids (or simple lipids), being less than 40%. This lecithin is extracted from various animal (egg yolk) or vegetable (soybean or others) sources.

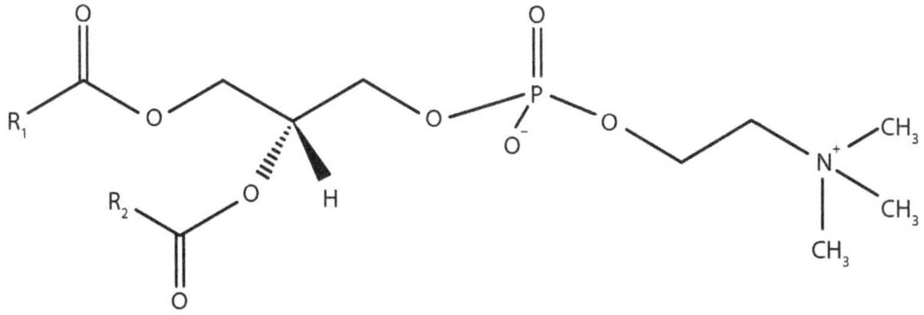

Figure 7.10 Phosphatidylcholine (R1 and R2 are the fatty acid carbon chains).

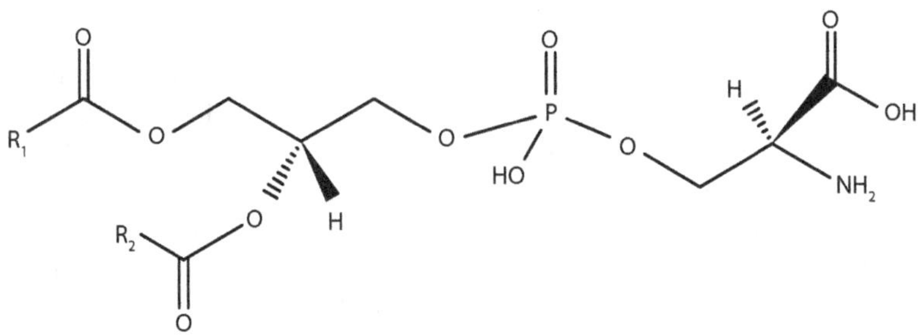

Figure 7.11 Phosphatidylserine (R1 and R2 are the fatty acid carbon chains).

Crude soybean oil is the vegetable oil with the highest lecithin concentration (1%–3%).

Choline is the most specific component of phosphatidylcholine and must be supplied to the body at a suitable level (400 mg daily for adults, EFSA source). A dietary supplementation in phosphatidylcholine is recommended in various pathological situations, but the role of choline has not been well discriminated from that of the remainder of the molecule.

7.7.2 *Phosphatidylserine*

Phosphatidylserine (Figure 7.11) is a minor constituent of cell membranes, but the nervous system, especially the white matter, contains large amounts of this phospholipid (up to 18%). One of its features is to have a very unsaturated fatty acid (often DHA) in *sn*-2 (R2). It can be considered a bioactive lipid, because it is involved in several physiological mechanisms, such as activation of protein kinase C and initiation of blood coagulation.

7.8 Apolipoprotein E

The apolipoprotein E (ApoE) forms a class of apolipoproteins found in the blood at the level of chylomicrons and mainly at the level of the very-low-density lipoproteins. They are able to bind specifically to receptors located primarily on the hepatocyte surface. These proteins are essential for the transport and the metabolism of the constituents of triglyceride-rich lipoproteins. They are, for example, essential for the cholesterol supply of nerve cells (internalization phenomenon). The genetic heterogeneity of ApoE was highlighted early. In humans, the coding gene for these proteins is located on chromosome 19.

Three alleles (isoforms) have been described: ApoE2, ApoE3, and ApoE4, thus determining six genotypes. With a frequency of 55%, the E3/E3 genotype is the most common in France; it is not linked to any known pathology.

The E4 allele is recognized as the main genetic risk factor for the non-familial form of Alzheimer's disease, especially if the subjects have two copies of that allele (E4/E4). Indeed, the risk of developing the disease for the carriers having the heterozygous form E3/E4 is 3.2 times higher than that of E3/E3, with the homozygote carriers (E4/E4) having a 11.6 times higher risk. It has also been suggested that the ApoE4 allele limits the transport of DHA in the brain, probably at the level of the blood–brain barrier. Thus, it would reduce the tissue metabolism of DHA and could cause very damaging local deficiency.

7.9 Evaluation of cognitive performances

The estimation of brain function is done in various ways, but in leaving aside the personality assessment, only the cognitive performances and the tests used to evaluate the efficiency of the various types of memory or the intellectual level are summarized here.

7.9.1 Different types of memory

Memory is, with learning, one of the higher features enabling humans to acquire and retain information deemed necessary for later use, thereby enabling the development of new skills gradually throughout life. Therefore, memory may be considered as the result of learning: it is the record of our contacts with the world that are saved in our neural network. In fact, despite great progress, memory remains at the center of much controversy for neuropsychologists. The identification and the systematic classification of the various types of memory, both in humans and animals, rely heavily on the observations of specific memory impairments after a lesion of a defined brain structure. Historically, the starting point of the classification of the different types of memory was based on the description in 1957 of a patient, by Scoville and Milner. The man, aged 27 years, had normal intelligence, but a bilateral and total ablation of the anterior areas in the median temporal lobes. The aim of that surgical operation was to remove severe epilepsy crises. Postoperatively, the patient developed total anterograde amnesia (inability to permanently keep new information) and a retrograde amnesia, covering a 1- to 10-year period before the operation. His other cognitive abilities (social, verbal, emotional) were preserved; the patient retained a capacity to acquire and retain some information under certain circumstances.

The description of the preserved and lost memory capacities in this patient has been the source of modern memory classifications. Conventionally, function is divided mainly into two categories:

1. Working or short-term memory, which seemed preserved in the patient
2. Long-term memory, itself divided into two subclasses
   - 2a. Declarative memory (or explicit memory), the conscious, intentional recollection of information. It depends on the integrity of the median temporal lobe.
   - 2b. Nondeclarative memory (or implicit memory), the unconscious, unintentional form of memory (procedural memory).

7.9.1.1 Working or short-term memory: Baddeley model

The working memory concept was proposed in 1974 by Baddeley and Hitch. Working memory is a form of short-term memory essential for all the cognitive activities. It is based on the temporary storage of a limited amount of information, enabling associations with other events. It requires all the immediate attention of the subject, and it was found to be involved in the performance of many cognitive tasks, remaining in relation to the long-term memory systems.

Recently, the trend is to separate working memory and short-term memory. The latter allows for information to be remembered for up to 1 minute and can restore it within that period. This memory is used, for example, when it is necessary to restore in a definite order a series of elements (such as figures) that were previously given. Generally, it is possible to hold between five and nine elements (one keeps usually seven, sometimes called "magic number").

Working memory has been separated from the short-term memory owing to the development of new analytical techniques. It allows for cognitive treatments on previously stored items. It could be involved in reasoning processes such as reading, writing, and calculating. It could also play a role, for example, when it is necessary to list in descending order a series of numbers previously set in an ascending order. This type of memory would be widely used by an interpreter in the exercise of a simultaneous translation.

7.9.1.2 *Long-term memory*

Modern conceptions of memory refer usually to long-term memory and break it down into two main subclasses: declarative memory and a nondeclarative memory, primarily on the criterion of a conscious or not conscious recall of information, respectively. This classification is based on the psychological characteristics of the information to be stored and on the brain structures for which integrity is required for managing this information (encoding, storage, retrieval). Long-term memory is used to store significant events experienced throughout the life, but also to remember the meaning of words and manual skills. Its capacity seems unlimited; it may last a few minutes, days, months, or years, or even a lifetime. Unfortunately, its reliability tends to decrease with age.

2a. **Declarative memory** (explicit or relational or recognition memory) refers to what is commonly considered the "memory" in the everyday life, and it corresponds to "to know that." It relates to the acquisition of specific autobiographical facts or general rules learned by repetition (learning by heart), such as grammar rules and multiplication tables. Then, these data consciously emerge expressing this memory through language; hence, the term declarative.

Neuropsychologists often use the term explicit memory. It has been subdivided into two categories: **episodic memory** that stores specific personal experiences and **semantic memory** that stores factual information.

- **Episodic memory** concerns autobiographical memories and allows explicitly to recall personal events stored in the hippocampus. It is in that brain area, a part of the limbic system, that different aspects influencing our behavior are managed: fear, aggression, and pleasure. It was shown that the hippocampus of London taxi drivers is overdeveloped related to their ability to connect places between them owing to a mental mapping. Recent investigations state that memories are stored for a few days in the hippocampus and then transferred slowly, during a few weeks, toward cortical areas specialized for the "older" memories. Episodic memory enables us to remember the past (e.g., meal of the day before, name of an old friend, or a historical date) and to plan for the next day. It is that form of memory that helps us to "travel mentally in space and time." A disorder of the episodic memory is observed in the Korsakoff's syndrome and during bitemporal amnesia.

- **Semantic memory** is the memory of knowledge and learned facts, such as those acquired during schooling (depending on the parahippocampal region). This memory allows, through a wide association network, to form the world's knowledge in a broader sense. It forms a database owned by everybody and that is quickly available, as in the memory of the meaning of words, capital names, traditional customs, object functions, color, or even the smells of objects. The semantic memory is used daily (e.g., highway code, grammar and conjugation rules). The disorders of this memory are associated with deficits in higher cognitive functions such as the language, the ability to recognize a face or perceive the shape of an object (gnosis), and the coordination of gestural activity (praxis). This is the memory affected very early in patients with Alzheimer's disease. The exact location of semantic memory in the brain, the conditions of its formation, and its links with the episodic memory continue to be debated.

2b. **Nondeclarative memory** (implicit or procedural memory) is a heterogeneous set of capabilities dependent of multiple brain systems; it corresponds to "to know how." Unlike declarative memory, it is inaccessible to consciousness. Neuropsychologists often use the term implicit memory. It is used during the motor, perceptual, or cognitive learning requiring repeated training, and it is mainly the memory of habits and skills, such as cycling. It is also subdivided into several subgroups, but the most important form is the form concerning the procedural memory. The latter, also known as automation memory, concerns the learning of many skill types. It operates at an automatic level and can be expressed in a normal manner without requiring the integrity of hippocampal structures, but it is dependent on the striatum. For example, the act of driving a car requires some automation related to procedural memory, allowing the driver not to be aware constantly of the necessary gestures. This memory is especially requested by athletes and artists.

7.9.2 *Evaluation of memory capacity by MMSE*

The Mini-Mental State Examination (MMSE, or Folstein test), commonly used in neurology, evaluates quickly memory problems reported by a subject. It allows to quickly give a physician or psychologist an idea of the condition of a patient with suspected dementia. The MMSE is a global cognitive evaluation tool and is used to establish various aspects of the cognitive state of patients, but not enough for a definitive diagnosis. It is only a tool to direct toward other investigations. Thus, it explores a person's sense of time and space, attention to details, ability to calculate, language skills, and constructional praxis. The results can be interpreted based on normative values, taking into account the age and education of the subject.

Below is an example of various tests, but there are variations depending on neurology centers.

Orientation test

1. What is the year?
2. What is the season?
3. What is the month?
4. What is the day?
5. What is the day of the week?
6. Where are we now?
7. In what state?
8. In what city?
9. What is the hospital's name?
10. What floor (or place, cabinet, etc.)?

Learning test

The examiner names three unrelated objects (e.g., flower, door, car) clearly and slowly and then the patient is asked to name all three of them and to remember them. The patient's response is used for scoring:

11. flower
12. door
13. car

Attention test and calculation

The examiner asks the patient to count backward from 100 by sevens (ask for five subtractions).

14. 100 – 7 = 93
15. 93 – 7 = 86
16. 86 – 7 = 79
17. 79 – 7 = 72
18. 72 – 7 = 65

As an alternative, ask to spell the word "WORLD" backwards? —> DLROW (one point per letter given in the correct order.)

Recall test

What were the three names of objects presented below?

19. flower
20. door
21. car

Test of language

22. Show a pencil: "What is the name of this object?"
23. Show a watch: "What is the name of this object?"
24. "Repeat after me: No if, and or but."
25. "Take the paper in your right hand." (The examiner gives the patient a piece of blank paper.)
26. "Fold it in half"
27. "Put it on the floor"
28. Show on a sheet of paper on which is written in large letters CLOSE EYES and say "Please read this and do what it says."
29. Give a sheet of paper and ask the following: "Make up and write a sentence about anything." (This sentence must contain a noun and a verb.)

Test of constructive praxis

30. Give the subject a piece of paper with a geometric design (two pentagons, 10 angles, but two must intersect) and request, "Please copy this picture."

Total score (from 0 to 30, one correct answer per point). Below 27 points, it can be considered that there is a deficit that may or may not progress to dementia. According to a recommendation of the French High Authority for Health, a score less than or equal to 24 points enables to evoke an altered state of consciousness and gives an indication to the diagnosis of dementia.

7.9.3 *Test of executive functions*

Executive functions, sometimes called "cognitive control" or "supervisory attentional system," are necessary for monitoring and execution of complex, new, and not automatic tasks. They allow the subject to set goals and to forecast a strategy to achieve them. These functions have been defined in various ways, but traditionally they are related to planning and execution behavior, abstract reasoning and judgments. From a practical point of view these higher level functions correspond to the efficiency that an individual displays to use his knowledge necessary to lead his everyday life. Although nonexclusive, the structures of the frontal lobes are considered as fundamental to the smooth running of these functions. Suitable techniques enable to appreciate one important aspect of executive functions and cognitive flexibility. The latter may be defined as the ability of an individual to adapt to new situations or new rules.

If a neurodegenerative disease is suspected, the neuropsychological examination should help to assess the extent and severity of the cognitive impairment by practicing a test of executive functions. Among the tests used in that field, the Trail Making Test (TMT) is the most used. Developed in 1944 by the US Army, the TMT has been very successful with clinical psychologists, due to its simplicity and speed. It is a motor and visual screening test involving attention and reaction speed.

The TMT has two parts:

- Part A is to connect with a pencil, as fast as possible and in ascending order, the series of numbers ranging from 1 to 25 semirandomly distributed on a sheet of paper. This part A is particularly indicative of the cognitive processing speed.
- Part B is to juggle between two alternating series: a series of figures and a series of letters. Thirteen circles containing the numbers from 1 to 13 and 12 circles with the letters A to L are distributed semirandomly on a paper sheet. The subject must connect as quickly as possible numbers and letters in their respective order, alternating each time between a number and a letter. This part B is useful to examine the executive functioning itself. Errors are recorded and also the time it takes the patient to perform the task.

Numerous studies have reported the existence of considerably reduced performances in various dementia syndromes such as Alzheimer's disease and even in cases of mild cognitive decline.

For screening Alzheimer's disease, the patient's opinion may be taken into account, but it may be distorted due to memory problems. The evaluation by a clinician is more objective, but it may be supplemented by questioning the immediate entourage, allowing an appreciation of the subject behavior in the activities of daily living. As emphasized by all the Alzheimer's associations, the entourage can efficiently contribute to the detection of the first signs of the disease. That approach helps the specialist to practice, if necessary, a medical evaluation, a brain imaging investigation, or neuropsychological tests. The latter particularly notes the changes in habits and everyday repeated incidents.

The following indices are particularly noted:

- Difficulties recording information
- Changes in mood and behavior
- Language problems

- Disorientation in time and space
- Judgment and reasoning losses
- Difficulty in managing money

The main purpose of the cognitive assessment by the clinician is the screening and the estimation of the severity of the trouble and its evolution. The forms of the disease where deterioration seems evident (MMSE < 20) are explored with more accurate and sensitive neuropsychological tools.

A large number of tests are currently available; 18 tests using a computer were already listed in 2008 (Wild et al. 2008, Status of computerized cognitive testing in aging: a systematic review. *Alzheimer's Dement.* 4:428–437).

The assessment of cognitive functions is often done with the Alzheimer's Disease Assessment Scale-cognitive (ADAS-cog) and may be performed in 45 min. This test was developed in the United States by Ros in 1984. Its score ranges from 0 to 70. It is thus possible to estimate the average degradation of the cognitive performances for the mild and moderate stages of the disease by a loss of 6–8 points per year against 2–4 MMSE points per year. Below, a summary of the various steps performed during this test is given. In an interview, a series of questions is asked about classmates and friends. Depending on the interview, the following are noted:

- Intelligibility of the spoken language.
- Comprehension.
- Lack of words.
- Reminder words (from a list of 10).
- Designate the fingers one by one in a given order.
- Set a date and time.
- Execution of orders, e.g., make a fist, double-tap with two fingers on each of your shoulders, keeping your eyes closed.
- Praxis: Fold a sheet of paper, put it in an envelope, stick the envelope, write the address, and apply a stamp.
- Constructive praxis: Copy simple geometric figures.
- Word recognition: 12 words among 24 words presented to the patient.
- Instructions reminder: Evaluation of the patient's ability to retain the instructions of the previous test.

This test can be completed by the assessment of noncognitive functions for depression, anxiety, autonomic dysfunction, behavior disorders, initiative in daily activities, psychotic symptoms, motor activity, restlessness, and night confusion.

7.9.4 Global deterioration scale

This scale, also called the scale of Reisberg, allows healthcare professionals to measure the progression of Alzheimer's disease. It sets out seven stages of deterioration of neuropsychological abilities during the disease development:

1. No cognitive impairment: no difficulty in daily life
2. Very mild cognitive deficit: forgets names and locations of objects; may have some difficulty finding words
3. Mild cognitive impairment: difficulty to find way in an unfamiliar place; difficulty in functioning at work
4. Moderate cognitive impairment: difficulty performing complex tasks (e.g., finance, purchasing, planning a meal)
5. Relatively severe cognitive impairment: needs help choosing clothes, needs to be reminded to go to the toilet
6. Severe cognitive impairment: loses sense of experiences and recent events from life; needs help with bathing or is afraid to bathe; needs increasingly help to go to the toilet or is incontinent
7. Very severe cognitive deficit: use a very limited vocabulary that will be reduced to only some words; loses the ability to walk and sit; needs help to eat
8. A state of dementia is often considered to be characterized from the stage 4.

7.9.5 Peabody picture vocabulary test

This test was designed to provide a quick estimate of language skills and the ability to pursue an education. It was developed in 1959 by L. M. Dunn. The examiner presents to the subject a series of images grouped by four on a page. These images are numbered. The word describing one of images is pronounced and the subject is asked to tell (or show) the corresponding number. The final score can be expressed as a percentage, mental age, or IQ.

7.9.6 Wechsler intelligence test

This test was created by D. Wechsler for children aged 6–17 years; it has been also adapted for adults. The most used version is that of 1996. It is the most-used psychometric test in the world and is used even for kids in major difficulties. It comprises two groups of questions in two general areas: verbal scales and performance scales. Verbal scales measure general knowledge, language, reasoning, and memory capabilities. Performance scales measure the abilities to solve orientation problems in space and analysis. This scale seems to offer certain advantages over the McCarthy Scales of Children's Abilities and also the Stanford–Binet Intelligence Scales.

The test is performed by an examiner using a comprehensive set of educational materials for a period of about 1 h. Results are usually processed in IQ. Scores above 130 characterize excellent; between 120 and 130 very high; and between 110 and 120 normal, but intelligent. From 90 to 110, the subject is medium and from 85 to 90 average.

7.9.7 Bayley motor test

The scale of development according to N. Bayley was established in 1969 to appreciate in children 0–3 years old the evolution of motor movements, cognition (with language), and behavior. The test runs according to age in the form of games for 15–60 min. The scores are calculated taking into account the questionnaires prepared by the parents and others in the immediate environment (childcare worker, supervisor). These scores are then compared to values established in many children of the same age having a development considered as a reference.

The motor movements are estimated by assessing the degree of control of the body, muscle coordination, manipulation of objects (agility of the hands and fingers), postural imitation, and recognition of objects by the sense of touch.

Mental development is evaluated in the areas of perception, memory, problem-solving, verbal communication, and number appreciation. The behavior is evaluated with tests of attention, orientation, awakening, emotion regulation, and quality of movements.

7.10 Kaufman children intelligence test

The Kaufman Test, known as the K-ABC, was established and calibrated in 1983 in the United States and 10 years later in France. Its author is a professor of psychology at Yale University, New Haven, Connecticut. This test is frequently used in academic or clinical psychology to estimate the intellectual level of children between 2.5 and 12.5 years old. It aims to measure intelligence and knowledge, while focusing more on the mental processes that on content. Basically, that test tries to determine how the child solves problems and his or her ability to do so by relying as little as possible on language, information, and skills.

The Kaufman Test includes several different scales of intelligence. In general, only some of them are used in one test.

- The *sequential processing scale* measures the ability of a child to solve problems dealing mentally with the stimuli within a series. For example, to reproduce a sequence of different hand positions, or a repetition in the order of a sequence of numbers given verbally by the experimenter.
- The *simultaneous processing scale* assesses the ability to solve problems requiring the organization and the integration of many

stimuli, in parallel or simultaneously. For example, name or describe an incomplete stylized drawing after its presentation; assemble several identical triangles to reproduce an abstract model; memorize the location of randomly placed images on a page.

- The *scale of knowledge* enables to evaluate a wealth of knowledge and skills acquired at school or through some awakening to the environment. For example, vocabulary; characters and known places; and arithmetic, reading, and understanding. One benefit of the K-ABC test is the possibility to use it with children suffering hearing disabilities, speech disorder, and language disorders, while visually impaired children are penalized.

7.11 Estimation of depression

In the area of the depression, the first scale in 20 points was established in 1977 in the United States by L. S. Radloff. It is still used today by researchers. Practically, other more simple scales may be used, including a scale developed by the Danish psychiatrist Per Bech for WHO, under the form of the Major Depression Inventory (MDI) questionnaire. This test is also known as the Center for Epidemiologic Studies Depression (CES-D) scale (http://www.therapiebreve.be/plus/tests/depression-mdi). That scale allows a self-assessment of the severity of a mood disorder, but it can be established in the presence of a specialist or remotely, by phone or even by mail. It enables the exploration of the nine symptoms identified as diagnostic criteria of a major depressive episode described in the Diagnostic and Statistical Manual of Mental Disorders (DSM-IV).

This 12-point questionnaire is given below:

1. Have you felt low in spirits or sad?
2. Have you lost interest in your daily activities?
3. Have you felt lacking in energy and strength?
4. Have you felt less self-confident?
5. Have you had a bad conscience or feelings of guilt?
6. Have you felt that life wasn't worth living?
7. Have you had difficulty in concentrating? (e.g., for watching TV, reading newspaper)
8. Have you felt very restless?
9. Have you felt subdued or slowed down?
10. Have you had trouble sleeping at night?
11. Have you suffered from reduced appetite?
12. Have you suffered from increased appetite?

Each point is the subject of a choice of an occurrence frequency: at no time, some of the time, slightly less than half the time, slightly more than half the time, most of the time, or all the time. The score calculation is weighted for each question according to the choice of the frequency.

Many other derivatives questionnaires derived from that of Radloff may be found on websites, but in all cases their use should be only taken as an alert for the examiner. Confirmation by a specialist consultation is always necessary.

7.12 Evaluation of aggressiveness and violence

The aggressive state has been studied for a long time in both children and adults. This is a relatively stable personality characteristic, because it has been shown that an aggressive state evidenced in children is very predictive of disorders and even delinquency in adolescence, often followed in adults by criminal behavior. Among the many methods of evaluation, the Social Dysfunction and Aggression Scale (SDAS) is frequently used.

The questionnaire is divided into 10 parts, with each part being the subject of a question and a note: 0 = absent, 1 = equivocal, 2 = poor, 3 = moderate, 4 = severe.

The assessments noted by the specialist include the following:

1. Irritability (estimates the reduced ability to cope with situations considered as provocative by the patient, anger, impatience, and the reduced ability to control responses)
2. Negative and uncooperative behavior (estimates the reduced ability to cooperate or to conform to a group)
3. Irritable or dysphoric mood (covers a type of unpleasant feeling where the patient is moody, testy, and fed up)
4. Socially disruptive and provocative behavior (estimates how the patient behaves in a provocative manner toward others, including sexually provocative behavior)
5. Nondirected verbal or vocal aggressiveness (verbal aggressiveness or noises assumed to represent aggression directed toward others or things in general)
6. Directed verbal or vocal aggressiveness (covers aggressiveness directed toward defined persons)
7. Physical violence toward things
8. Physical violence toward staff
9. Physical violence toward others that staff
10. Self-mutilation

Another also widely used method is the Aggression Questionnaire established by C. Meesters in 1996. The protocol described by this author is given below. The questionnaire is divided into four parts: physical aggression, verbal aggression, anger, and hostility. The subject must note his or her answers on a scale from 1 (yes, that is true) to 5 (no, that is wrong). The final note is between 29 and 145.

Physical aggression

1. Once in a while I can't control the urge to strike another person
2. Given enough provocation, I may hit another person
3. If somebody hits me, I hit back
4. I get into fights a little more than the average person
5. If I have to resort to violence to protect my rights, I will
6. There are people who pushed me so far that we came to blows
7. I can think of no good reason for ever hitting a person
8. I have threatened people I know
9. I have become so mad that I have broken things

Verbal aggression

10. I tell my friends openly when I disagree with them
11. I often find myself disagreeing with people
12. When people annoy me, I may tell them what I think of them
13. I can't help getting into arguments when people disagree with me
14. My friends say that I am somewhat argumentative

Anger

15. I flare up quickly but get over it quickly
16. When frustrated, I let my irritation show
17. I sometimes feel like a powder keg ready to explode
18. I am an even-tempered person
19. Some of my friends think I am a hothead
20. Sometimes I fly off the handle for no good reason
21. I have trouble controlling my temper

Hostility

22. I am sometimes eaten up with jealousy
23. At times I feel l have gotten a raw deal out of life
24. Other people always seem to get the breaks
25. I wonder why sometimes I feel so bitter about things
26. I know that my "friends" talk about me behind my back
27. I am suspicious of overly friendly strangers
28. I sometimes feel that people are laughing at me behind my back
29. When people are nice, I wonder what they want

Index